T0329691

ADVANCED
CHIPLESS RFID

WILEY SERIES IN MICROWAVE AND OPTICAL ENGINEERING

Editor: Professor Kai Chang, Texas A&M University

The Wiley Series in Microwave and Optical Engineering publishes authoritative treatments of foundational areas central to Microwave and Optical Engineering as well as research monographs in hot-topic emerging technology areas. The Series was founded in 1988 and to date includes over 100 titles.

A complete list of the titles in this series appears at the end of this volume.

ADVANCED CHIPLESS RFID

MIMO-Based Imaging at 60 GHz – ML Detection

By

NEMAI CHANDRA KARMAKAR
MOHAMMAD ZOMORRODI
CHAMATH DIVARATHNE

Library of Congress Cataloging-in-Publication Data

Names: Karmakar, Nemai Chandra, 1963– author. | Zomorrodi, Mohammad, 1973–
author. | Divarathne, Chamath, 1983– author.
Title: Advanced chipless RFID : MIMO-Based Imaging at 60 GHz–ML Detection / by Nemai
Chandra Karmakar, Mohammad Zomorrodi, Chamath Divarathne.
Description: Hoboken, New Jersey : John Wiley & Sons, Inc., [2016] | Series:
Wiley series in microwave and optical engineering ; 1187 | Includes
bibliographical references and index.
Identifiers: LCCN 2016014411 | ISBN 9781119227311 (cloth)
Subjects: LCSH: Radio frequency identification systems. | MIMO systems.
Classification: LCC TK6570.I34 K364 2016 | DDC 621.3841/92–dc23 LC record available at
https://lccn.loc.gov/2016014411

Printed in the United States of America

10 9 8 7 6 5 4 3 2 1

CONTENTS

PREFACE

The author's group has developed various chipless RFID tags and reader architectures at 2.45, 4–8, 24, and 60 GHz. These results were published extensively in the form of books, book chapters, refereed conference and journal articles, and finally, as patent applications. However, there is still room for improvement of chipless RFID systems. In this book, we proposed advanced techniques of chipless RFID systems that supersede their predecessors in signal processing, tag design, and reader architecture.

The book introduces a few novel and advanced-level high-capacity data encoding and throughput improvement techniques for fully printable multibit chipless RFID tags and reader systems, respectively. These techniques enhance data content capacity of tags and perform reliable tag detection for readers at the instrumentation, scientific, and medical (ISM) frequency bands 2.45, 24, and 60 GHz. First, a comprehensive review of existing chipless RFID tags provides the state of the art in the field and exposes impediments for commercial success. The limiting factors for commercialization of reported chipless RFID tags are (i) printing errors, (ii) degradation of tag performance on low-grade laminates, (iii) low data capacity, (iv) errors in tag reading in industrial environment, (v) reading reliability, and (vi) read range. This book addresses these limitations and provides solutions with an image-based tag design and advanced signal processing techniques.

The book provides the details of the new approaches – electromagnetic (EM) imaging, high-capacity data encoding, and robust tag detection techniques. In the introduction chapter first, a comprehensive review of the available and reported chipless RFID systems is presented. Then, their above-mentioned impediments for commercial success are analyzed. The analysis shows that the conventional techniques used for chipless RFID tag encoding and detection do not address the challenges imposed by commercial grade tags and reader systems. This encourages the researchers for new techniques and approaches in this field.

The book is divided into two main parts. Part I of the book, "EM Image-Based Chipless RFID System," introduces the novel EM imaging concept for data extraction from a 60-GHz chipless RFID tag. Part II "Advanced Tag Detection Techniques for Chipless RFID Systems" presents smart tag detection techniques for existing chipless RFID systems and an innovative MIMO-based tag detection technique for high content capacity and zero guard-band tag detection. These approaches have been fully developed and tested in Monash Microwave, Antenna RFID and Sensor Research Group (MMARS) at Monash University.

In Part I of the book, the fundamental of EM imaging at millimeter-wave band 60 GHz for data extraction is introduced followed by the EM imaging through synthetic aperture radar (SAR) technique. It is shown that the millimeter-wave EM imaging has significant potentials for commercialization of chipless RFID. The EM imaging technique exploits advantages of RFID systems including their flexible non-line-of-sight (NLoS) operation and high data capacity benefit. Moreover, the proposed EM imaging technique inherits low-cost advantages and fully printable features of the barcodes on low-grade packaging materials. The downside of the conventional SAR-based EM imaging technique, requirement for physical movement of the reader antenna, is addressed by the new idea of MIMO-SAR technique. With the proposed MIMO-based EM imaging, no relative movement of the reader and tag is required hence very fast tag imaging is achievable. Finally, the MIMO approach is optimized through global genetic approach for minimum hardware complexity and to introduce a complete solution for chipless RFID system. In this pursuit, the system elements and technical requirements are discussed in details. The proposed approach to the EM imaging technique enhances the content capacity of the chipless systems to a commercial level, for example, EPC Global Class 1 Generation 2 with 64 data bits.

The main emphasis for Part II of the book is to introduce a few new smart tag detection techniques for chipless RFID systems. Researchers were mainly focusing on improving the RFID reader architecture [1,2] and the chipless tag design in conventional approaches [3] and paying less attention to signal processing. As a result, most signal processing techniques being used in chipless RFID systems are primitive and should further be investigated. The first part of Part II focuses on advanced signal processing techniques that significantly improve the tag detection rate and tag reading range for the existing reader architecture [1] and tag design [4]. In addition, the proposed techniques allow removing the guard band presented in frequency-domain tags allowing the spectral efficiency to be improved. As a result, data capacity of the frequency-domain tags can be improved. Maximum-likelihood (ML)-based detection techniques have shown improved performances in communication systems compared to reported techniques such as threshold-based detection techniques [5]. The motivation for this work is to apply the ML detection techniques for chipless RFID tag detection so that the existing RFID system would perform better. One limitation of likelihood-based techniques is its exponential increase in computation complexity with higher number of data bits. Two computationally feasible tag detection techniques have been introduced to overcome this challenge. With these new tag detection techniques, computation complexity only increases linearly with the number of data bits. The second part of Part II presents a novel MIMO-based chipless RFID system and required tag detection techniques that can be used to improve the spectral efficiency, hence increasing data bit capacity.

We hope that the book will contribute significantly to the field of chipless RFID removing many practical barriers for commercialization of the technology.

NEMAI CHANDRA KARMAKAR
MOHAMMAD ZOMORRODI
CHAMATH DIVARATHNE
Melbourne
November 2015

REFERENCES

1. N.C. Karmakar, R. Koswatta, P. Kalansuriya and R. E-Azim, *Chipless RFID Reader Architecture*, Artech House Publishing, 2013.
2. R.V. Koswatta and N.C. Karmakar, "A Novel Reader Architecture Based on UWB Chirp Signal Interrogation for Multiresonator-Based Chipless RFID Tag Reading," *IEEE Transactions on Microwave Theory and Techniques*, vol. 60, no. 9, pp. 2925–2933, 2012.
3. R. Rezaiesarlak, and M. Manteghi, *Chipless RFID, Design Procedure and Detection Techniques*, Springer, USA, 2015.
4. S. Preradovic and N.C. Karmakar, *Multi-Resonator-Based Chipless RFID*, Springer, USA, 2012.
5. R. Koswatta and N.C. Karmakar, "Moving Average Filtering Technique for Signal Processing in Digital Section of UWB Chipless RFID Reader," *Proc. 2010 Asia Pacific Microwave Conference*, Yokohama, Japan, December 7–10, 2010.

ACKNOWLEDGMENT

The book is the outcome of two PhD level thesis works under the supervision of the first author. The PhD scholarship was supported by the Australian Research Council (ARC) Discovery Project grant DP110105606: Electronically Controlled Phased Array Antenna for Universal UHF RFID Applications. Therefore, the support from ARC is highly acknowledged. The editor-in-chief of Wiley Mr. Brett Kurzman, Editor, Global Research, Professional Practice and Learning, Wiley and Mr. Alex Castro, Senior Editorial Assistant of Wiley were very supportive from the inception of the book project to the end of production. Their support is highly acknowledged. The two student authors were cosupervised by Professors Jeff Walker and Jamie Evans of Monash University. Their valuable suggestions and technical assistance are also acknowledged. During the course of the research work, the team members of the authors' research group, Monash Microwave, Antenna, RFID, and Sensor (MMARS) Laboratory of Monash University, were very supportive to the PhD projects. The supports from the electronics and mechanical workshops of ECSE department of Monash University were instrumental for the research outcomes that are produced in the book.

The family members of the Authors had to endeavor their absence during this research. Their support and companionship are gratefully acknowledged.

NEMAI KARMAKAR
MOHAMMAD ZOMORRODI
CHAMATH DIVARATHNE
Melbourne
November 2015

PART I

EM IMAGE-BASED CHIPLESS RFID SYSTEM

1

INTRODUCTION

The area of contactless identification systems is growing rapidly into a multi-billion dollar market. It covers a broad range of applications including supply chain management, manufacturing, and distribution services. Examples of these applications include consumer packaged goods, postal items, drugs, books, airbag management, animal tracking, pharmaceuticals, waste disposal, clothes, defense, smart tickets, people tracking such as prisoners, hospital patients, patients in care homes, and leisure visitors as shown in Figure 1.1. Tough trading conditions due to the global competition strive industries to attain more process efficiencies. Therefore, effective goods tracking systems are required to assist the implementation of the modern management system.

In general terms, any application that involves object identification, tracking, navigation, or surveillance would benefit from an identification system. Several hundred billion tags per year are required by this wide area of applications [1].

In this market, every application has its own technical and financial specifications. Main applications, those that need a huge number of tags, require high data encoding capacity and survive only with a very cheap tag solution. For others, secure identification and antitheft tagging is more important. In some cases, the tag size is a key factor and for some others proper identification of highly reflective items such as liquid containers or metal objects is of more importance. Reading range would be another imperative factor for many applications.

Advanced Chipless RFID: MIMO-Based Imaging at 60 GHz–ML Detection, First Edition.
Nemai Chandra Karmakar, Mohammad Zomorrodi, and Chamath Divarathne.
© 2016 John Wiley & Sons, Inc. Published 2016 by John Wiley & Sons, Inc.

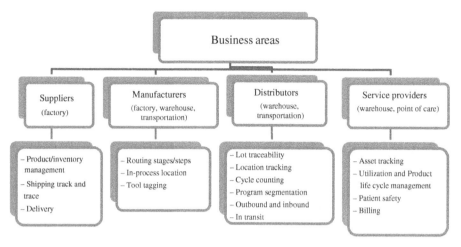

Figure 1.1   Application areas of identification systems.

Irrespective of all priorities, there are two main factors that significantly matter in all applications: the *data encoding capacity* and the *system cost*. For applications with millions of items for tagging, high data capacity of the identification system is a must. For applications with a limited number of objects, high data encoding capacity would be also beneficial to secure the identification process or provide higher reading reliability by sacrificing some of the available bits. The cost reduction is the main initiative for the usage of identification systems in industry; hence, the cost of the identification system and its tagging price must be low and competitive enough to initiate the request for the system. Otherwise, there would be no demand for such systems.

The cost of identification systems, like any other broadcasting service, has two parts: the reader and the tag. The reader cost is normally a fixed cost irrespective of the number of tags. However, the price of the tag attached to every individual item is the most costly part of the whole system. Specifically when the number of items is in the order of millions, the tag price plays a major role in the system's total cost. For such applications, a tag price of only $1 would increase the total cost of the system to a level that restricts the usage of identification systems. Therefore, the tag price should be kept as small as possible to offer a reasonably low identification system cost.

1.1   BARCODES AS IDENTIFICATION TECHNOLOGY

Barcode is an optical-based, machine-readable technique for identification purposes. It has been established in various industries for many decades with proven applicability. Barcode provides an *extremely low-cost* solution for identification of items to which it attaches. Originally, barcode are comprised of many parallel printed dark lines. The tag's data are systematically represented by varying the widths and

Figure 1.2 Data encoding limitation of a 1D barcode tag due to diffraction effect.

spacing of those parallel lines. This type of barcode, dominant in many applications, is normally referred to as linear or one-dimensional (1D) barcode. Data encoding capacity of the barcode tag is restricted by the diffraction of light through the edges of the lines, the reader sensitivity, and the reading distance, as shown in Figure 1.2. Diffraction restricts the minimum detectable line width as well as the minimum distance between two adjacent lines. This means that for increasing the data encoding capacity of the barcode, the only way is to increase the length of the tag. As the data encoding capacity of barcodes is proportional to the tag's size, it may result in an unreasonable tag size for many applications. This issue is considered as the main limitation of the barcode systems. The 1D barcodes have evolved into rectangles, circles, dots, hexagons and other two-dimensional (2D) geometric patterns to enhance the data encoding capacity. This has resulted in new machine-readable optical labels known as quick response (QR) code. QR codes use four standardized encoding modes to efficiently store data. The maximum storage capacity of QR codes can be up to 7000 characters, which is better than that of barcodes [2]. However, barcodes and QR have many operational limitations. They are very labor intensive as every tag needs to be read/scanned individually. Moreover, being an optical-based system, a clear line-of-sight (LoS), known as optical LoS, is also necessary for proper reading. This means that the tag shall be always printed and exposed on the products and the scanner requires clear optical LoS to read the barcodes or QR codes. Barcodes inside clear polyethylene bags cannot be read due to the light reflection of the bags. Any damage or dirt on the barcode results in improper reading. The reading distance between the optical scanner and the tag is also limited when considering the light dispersion/attenuation in free space and diffraction effect on the tag surface. Normal reading distance in optical systems is limited to few centimeters. Moreover, barcode is not a secure means of communication as tags can be easily reproduced by a cheap inkjet printer. The reading errors of barcodes depend on applications and many industries lose billions of dollars as compensations and damages each year. For example, optical barcode-based luggage handling has approximately 20%

reading errors and airlines are paying more than $2 billion/year as compensations to passengers.

To address positive aspects of barcodes, no doubt a very cheap tag solution and proven applicability in identification systems are the most important factors. Its few cents tagging solution is very attractive for many applications, specifically for industries with millions of products. Being accepted globally for almost half a century also provides it a unique superior opportunity that makes it very difficult for other technologies to compete. The globally accepted international barcode quality specification standards, ISO/IEC-15416 (linear) and ISO/IEC 15415 (2-D) [3], and no privacy issues involved with the barcodes usage are highly regarded by many users.

Moreover, barcode systems provide a fairly good reading accuracy that is almost comparable with what other new techniques are offering [3]. Another good aspect of the barcode is that the accuracy of the reading process is almost independent of the items on which tags are placed.

1.2   RFID SYSTEMS

The usage of light waves as communication mean in the barcode systems causes many technical and operational limitations as discussed before. As an alternative approach, the use of EM waves for identification and tracking of objects was first proposed by Watson-Watt in 1935 [4] and coined as the radio frequency identification (RFID) system. In an RFID system, the reader sends an electromagnetic (EM)-wave interrogating signal toward the tag. This signal is then processed by the tag's microchip unit and backscatters the signal toward the reader. This backscattered signal carries the tag identification information and is received and processed by the reader to retrieve the data.

Figure 1.3 shows the generic configuration of the RFID system. As the EM wave is not obstructed by barriers, the system does not need a LoS link between the reader and tag. This provides a number of opportunities for an RFID system. For example, the tag may hide inside the item and not necessarily be exposed on the object as the barcode system does. Moreover, many reader antennas are omnidirectional; hence, they can detect tags irrespective of their position with respect to the reader. Multiple

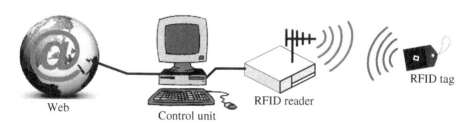

Figure 1.3   RFID general system structure.

tag reading is also feasible in an RFID system, bulk detection scenario. The RFID reading distance may be much greater than that of barcodes as the EM waves are much less attenuated in free space than light waves. The more attractive part of an RFID system is its higher data encoding capacity, which is not comparable to the barcode, as the data are encoded by a microchip. Moreover, many security codes can be easily manipulated inside the microchip to provide more secure communication.

1.3   BARCODES VERSUS RFID

Optical-based identification systems, barcodes and QR codes, and RFID systems all have their own advantages and limitations. This means that each system would be suitable for different purposes and under different circumstances. Although the majority of users still consider barcode systems as the most cost-effective way to handle the circulation and inventory management of equipment, the indication of changing market occurred in 2003, when Walmart adopted and mandated RFID tagging for all its suppliers. Walmart's motto of mandating RFID is to obtain seamless information from the manufacturing point to the ends of sales when the goods are sold and the boxes are crashed. There are numerous discussions and studies in industry and academia about the suitability of these three systems [5–9]. It is almost agreed that there is no clear superiority of one technology over others when both the cost and operational flexibility are considered simultaneously. In general, it is upon each specific industry to select the most suitable technology based on their needs and budgets. The benefit of barcode technology comes from their low-cost implementation. It is well established in industry and has fairly enough content capacity for many industrial and commercial applications. QR codes offer higher data encoding capacity while working almost on the same basis as barcodes. RFIDs are popular and more appropriate technology than barcodes for many industries as it can provide higher content capacity and much more operational flexibilities. For example, Cisco recently announced its new idea of Internet-of-everything (IoE) based on RFID systems [10]. However, conventional RFID systems are associated with limitations too. The main issue for RFID is that many industries cannot afford the system cost. To mitigate the potentials of RFID systems to compete with optical barcodes and being accepted by more applications, it is required to reduce the cost of the RFID tag to a level similar to optical barcodes, say less than a cent. A fully printable tag that is still able to provide the same or higher data encoding capacity compared to 1D barcodes with more operational flexibility would be highly welcomed by industries. A fair comparison among barcode, QR codes, and RFID systems is provided in Table 1.1.

1.4   CHIPLESS RFID TAG FOR LOW-COST ITEM TAGGING

As addressed in the previous section, RFID systems show many technical and operational superiorities over barcodes and, therefore, are suggested as the most promising technique for barcode replacement. However, to date, this has not

TABLE 1.1   Barcode, QR codes, and RFID [8,9]

	Data Capacity (max)	Tag Cost (¢)	Reading Speed	Unique Advantage	Operational Limitation	Reading Distance
Barcodes	20 bits	0.5–1	Relatively quick	Cheap and accurate	Optical LoS	Few centimeters
QR codes	7000 characters	1	Relatively quick (depends on device)	Versatile	Optical LoS	Few centimeters
RFID	4 million characters	>30	Very fast	Many technical superiorities	EM LoS or NLoS	Tens of meters

happened for main applications with billions of yearly tag requirements because of the higher cost of the RFID system. The RFID system cost mainly depends on its tag expense like any other broadcasting system. The total cost of identification system is mainly governed by the tag's cost only when the tag number is significant. Normally, the reader system does not contribute significantly in operational cost as it is a fixed cost.

System cost = fixed cost + N × tag cost

N: number of tags; fixed cost = reader electronic + middleware costs

if $N \to \infty$

then : total system cost $\approx$ tag cost × N (1.1)

In RFID systems, every tag needs a silicon chip to encode data. This results in an RFID tag cost that is many times more expensive than the barcodes. Significant investments and research have been spent on lowering the price of microchips to less than a cent and thus make it comparable to that of barcodes. However, the application-specific integrated circuit (ASIC or IC for short) design and testing along with the tag antenna and ASIC assembly still result in a costly manufacturing process [11–13]. Furthermore, as the price of every silicon chip directly depends on its size on the wafer, the minimum predictable cost of an RFID chip with the quantity of billion cannot be less than 5¢, which is still not competitive with 1¢ tag price of barcodes [11,14]. Despite all recent improvements in silicon chip technology, silicon chips remain too expensive to be part of every RFID tag [15]. Considering the minimum predictable cost of a chipped tag, the total cost of an RFID system in applications with millions of tagging requirements would be much higher than that of barcodes. Therefore, the price of chipped RFID tags remains as the first and foremost hurdle for their deployment in applications with low product costs, for example, groceries. Based on the company's potential and system affordability, the relation between the tag cost and its volume is shown in Figure 1.4 [1]. Based

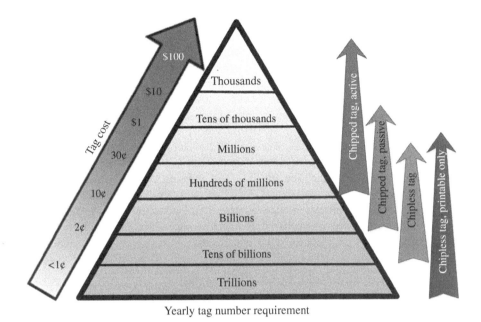

Figure 1.4 Expected tag volume versus tag cost. *Source*: IDTechEx [1].

on this model, the RFID tag cost would be a large barrier in organizations with high tagging requirements. Therefore, RFID tags without a chip, named as chipless tags, appear to be necessary for the commercialization of RFID systems in main applications with billions of yearly tag requirements [1]. As shown in Figure 1.4, the chipless tags are necessary to satisfy the requirements of medium- to large-sized organizations, with a tag price to be down to 1¢. Although the chipless tags eliminate the need for the silicon chip, the most expensive element of the tag structure, there are other factors that may surge the tag cost. The tag fabrication process, the requirement of tag to be affixed with other costly elements rather silicon chip, and their installation procedure may elevate the tag cost higher than the targeted value of 1¢, radiofrequency surface acoustic wave (RFSAW), for example [16]. Therefore, based on the worldwide well-accepted model, the main stream industries with billions of yearly tag requirements will only be satisfied through a *fully printable chipless* tag structure. Any technique or suggested solution for a chipless RFID system must consider this critical point of the industries for a printed tag; otherwise, the proposed technique finds no place in the identification market or at least in main stream industries.

However, the way the data can be encoded in a fully printable chipless tag is a big challenge that opens a new area of research. To date, many techniques and approaches have been proposed for a data encoding scheme in a chipless tag structure [11,17–19]; however, very few products are available on the market.

1.5 CHIPLESS RFID SYSTEMS

Reducing the RFID tag price to below 1¢, printable chipless tags appear to be the only solution. There are many proposed techniques in the open literature on designing a chipless and passive tag structure with the mandated data encoding capability for identification purposes. This section mainly focuses on reviewing those techniques and approaches and exploring their potential advantages and limitations.

The communication between the RFID reader and the tag is accomplished through the use of EM waves. For the RFID chipless systems, the tag does not require any processor unit; hence, all the reading and coding processes are accomplished in the reader. The basis of the chipless system is that the reader receives the tag's backscattered signal in different domains and processes the signal in different domain to retrieve its encoded data. This leads to the time-domain-based, frequency-domain-based, phase-based, or hybrid systems [20–22]. Figure 1.5 shows the classifications of the chipless RFID systems that are reviewed in this section.

1.5.1 Time-Domain-Based Chipless RFID Systems

In a time-domain-based system, the reader interrogates the tag with a series of pulses [23,24]. The tag then retransmits the signal as a train of echoes with some time delays with the data encoded in the delayed responses. Manipulation of the delays can be handled directly on the EM waves domain. It is possible to convert the EM wave to another type of medium, acoustic wave, for example, and then delays are deployed in the signal. After manipulation of the data as delayed responses, the EM wave is retransmitted to the reader. When an alternative medium is used for data encoding purposes other than the EM wave, extra elements are needed for conversion [25]. For instance, in the surface acoustic wave (SAW) technique, the interdigital transducer element is used to convert an electromagnetic wave into a mechanical wave, which travels much slower than EM waves. This surface acoustic wave propagates through a piezoelectric element and then it is reflected back by a number of reflectors toward the reader. The requirements for extra elements in the SAW system elevate the tag

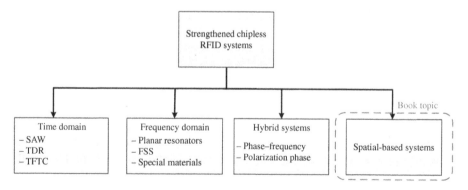

Figure 1.5 General classifications of chipless RFID systems.

price. Lowering the tag cost in time-domain-based systems mandates a printable tag structure without conversion of EM waves to other types of media. In this approach, the tag operates on the time-domain reflectometry (TDR) principle. The TDR tag normally consists of different types of transmission lines with multiple discontinuities [23,24,26–28]. Every discontinuity creates a reflection in the passing signal that shall be detected by the reader as the encoding technique. This approach provides the planar version of the tag structure; hence, a very low tag cost expectation through direct printing is claimed to be feasible. There are, however, some basic limitations that restrict their usage in real scenarios. Considering the much higher speed of EM waves than mechanical/acoustic waves, the required circuit size is remarkably large in creating detectable delays in the backscattered signal. For example, almost an $80 \times 30\,mm^2$ board size is required to encode only 4 bits of data, with the tag size rapidly increasing with a higher amount of data [24]. Moreover, the claimed 4-bit capacity is also based on the fabricated tag structure with PCB technology, and no information on the printed tag using conductive ink on paper were declared. The structure also includes some via holes that are not possible to mount on the commercial tag structure that is fully printable. In another work, the use of a transmission line of 2 m is reported to have a 4-bit coding capacity [29], which results in a tag size of $112 \times 53\,mm^2$ while the FR9151, Dupont was used as the substrate. The tag also includes the ground plane that increases the tag cost. No information was revealed on the performance of such a printed tag. There are some techniques proposed by other researchers to decrease the tag size in time-domain-based systems; however, the total performance of the system was significantly degraded [30,31].

In summary, it can be concluded that the time-domain-based systems have major limitations in tag cost reduction and on providing enough data encoding capacity in a reasonable tag size. Although SAW tag, the most successful chipless RFID product available in the market [32], is time domain based, it appears that the main application of RFID systems will not be solely satisfied by the time-domain-based approaches.

1.5.2 Frequency-Domain-Based Chipless RFID Systems

The reader in a frequency-domain-based RFID system normally interrogates the tag with a wide band signal. The interrogation signal is reflected back toward the reader, while specific resonances based on the tag structure are manipulated in the frequency domain of the backscattered signal. Alteration of the frequency domain of the interrogation signal is normally known as the tag's frequency signature. This process is figuratively shown in Figure 1.6. The reader then extracts the encoded data on the tag structure based on the detected resonances of the tag frequency signature [17,33,34]. To create specific resonances on the frequency domain of the interrogation signal, there are different types of approaches:

(i) Planar structure resonators [17]
(ii) Frequency-selective surface (FSS)-based resonators [35,36]
(iii) Special materials usage for resonant purposes [37–42].

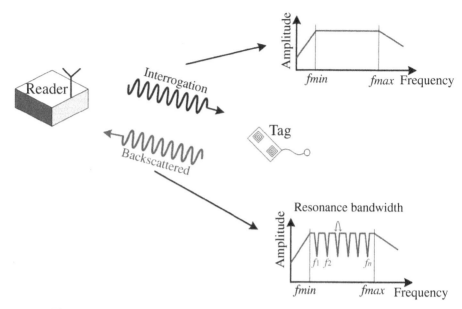

Figure 1.6 Working basis of the frequency-domain chipless RFID systems.

1.5.2.1 *Planar Structure Resonators*   Jalaly and Robertson [43] proposed some RF barcode-type structures for data encoding. The structure is composed of planar dipoles, similar to barcodes that resonate at different frequency bands. To enhance the content capacity of the tag, it was proposed to use all available industrial, scientific, and medical (ISM) radio frequency bands. Based on the proposal, every ISM band is capable of encoding a maximum of 5 bits. This expectation is based on simulation results. The precisely fabricated tag on the Taconic-TLY ($\varepsilon_r = 2.2$) substrate creates detectable resonances. However, based on the simulation results, the occurred resonances have 4–6 dB resonance deep that would be difficult for detection purposes in real scenarios. It is very obvious that the same structure is not capable of creating detectable resonances if the tag is printed on a lossy paper substrate with low conductive ink. Moreover, system reliability decreases when multipath and clutter interferences are considered. The effect of printing inaccuracy on the length of the dipoles, which creates a resonance shift, is also another factor that affects the consistency of the system performance.

Microstrip spiral resonators were proposed to create sharp resonances on the frequency domain of the backscattered signal [17]. The chipless tag also comprises a cross-polarized transmitting and receiving microstrip ultra-wideband (UWB) disk-loaded monopole antenna. It was shown that the system is capable of encoding up to 35 bits at 4 GHz frequency bandwidth. This is a significant success in the content capacity of the chipless RFID systems. However, certain circumstances were considered for the measured data capacity. First, the tag was fabricated through a costly process using the printed circuit board (PCB) technology on the low loss material of Taconic TLX-0 ($\varepsilon_r = 2.45$). Moreover, the printed tag with conductive ink on a paper base will experience significant Q-factor drop due to the printing errors, high loss of the paper, and low conductivity of the ink. These factors may limit the actual data encoding capacity of the proposed structure. However, the proposed idea in this work suggests a wide range of potentials for chipless RFID systems.

Fractal structures, as the load section of a wideband monopole antenna with encoding capability, were proposed by Balbin [44]. Based on the type and length of the fractal structures, multiple resonances would be detectable in the backscattered signal. However, due to the complex coupling behavior among the branches of a fractal structure, the resonance frequencies alteration seems to be unpredictable. This may restrict the actual usage of the fractal structure for the RFID applications. Moreover, in the proposed work, the behavior of precise fabricated structures was only considered and there was no study on the printable version of the fractal resonators. Considering the very tiny and complex shape of fractal structures, it is expected that printing issue may cause more complexity on the frequency resonances of the tag. No further results have been reported on the fractal resonators for the RFID applications.

Using split ring resonators (SRRs) was also proposed for the data encoding purposes on the chipless RFID systems [45]. The proposed tag structure was printed on the polycarbonate ($\varepsilon_r = 3.25$) material and measured on the frequency range 8–12 GHz. The tag was capable of encoding 4 bits of data on a 4 GHz spectrum bandwidth. The authors suggested using waveguide antennas to create a small illumination zone. By repeating the SRRs on the tag surface, more encoding capacity would be feasible. The proposed technique would be suitable for security of credit/personnel cards. However, due to the low encoding capacity, the proposed approach is not practical for identification purposes.

As an advanced approach, the usage of complex natural resonances was proposed to enhance the content capacity of a chipless RFID system [46]. Based on the singularity expansion method (SEM), complex natural resonances are aspect-independent parameters that include some structural information of the scattering target. This technique suggests encoding of up to 24 bits in a small area of 24×24 mm^2 that would be an extensive encoding enhancement compared with other available approaches in the chipless RFID systems. However, the system requires complex signal processing as multiple switching between time and frequency domains are required to decode the tag's content. Moreover, the performance of the printed tag was not shown in the proposed communication. The limited conductivity of the conductive ink and high loss tangent of the paper significantly changes the poles' positions of the tag responses; hence, the proposed data encoding scheme is challenged. In addition, the tag structure is very tolerance dependent as it comprises the line width and gaps in submillimeters (0.2–0.7 mm). This precise structure cannot be manipulated successfully through commercial printing facilities for low-cost tag production purposes. The effect of printing inaccuracy was not also considered in the proposed approach. However, the SEM approach opens up a new area of research for chipless RFID systems and may be combined with other conventional techniques for enhancing their potentials for massive commercialization.

1.5.2.2 Frequency-Selective Surface-Based Resonators

Frequency-selective surfaces (FSSs) are constructed as the rows and columns of a particular resonant structure designed to perform a (or a combination of) low-pass, band-pass, high-pass, and band-stop filtering functions on the incident plane wave passing through, in a measurement known as "transmittance" [43,47]. Although barcodes and RFID systems are reflection-based structures, sometimes the FSS are used for creating resonances in the reflection direction and hence are suitable for RFID applications.

Costa *et al.* [35] proposed the usage of high-impedance surface-based multiresonators. The structure is based on a finite metallic FSS comprising of several concentric square loop resonators. The reported tag structure occupies an area of $45 \times 45\,mm^2$ and is capable of encoding 5 bits of data content. Therefore, the data encoding density is $0.25\,bit/cm^2$, which is very low considering the requirements of the main RFID applications. The tags are readable from 55 cm distance with 0 dBm transmit power. The system had a good performance for different tag orientation scenarios. However, the system needs a calibration measurement process for proper reading to cancel out the effect of multipath interference and the effect of antennas; otherwise, the system fails to read and detect the tag's data. Moreover, all measured data in the published work is based on the fabricated tag structure with PCB standards on an FR4 substrate. No result based on the printed tag structure was revealed.

In another work, a stacked multilayer patch antennas is used as an all-pass network to provide more robustness with respect to multipath and clutter interferences [36]. In the proposed theory, instead of relying on an amplitude–frequency response of the multilayer structure, the phase–frequency response is considered. A three-layer structure with an area of $18 \times 18\,mm^2$ is able to decode 2 bits of data content based on the simulation results. No measurement of the proposed tag structure was shown. However, the performance of the proposed structure is directly linked to the conductivity and loss tangent of the substrate. This suggests that the printed tag may show significant performance degradation with respect to the simulation result. Moreover, printing of a three-layer structure increases the tag fabrication cost and may not meet the low-cost expectation of the RFID tag for the main industries.

1.5.2.3 Material-Based Resonators

CorssID [37], a telecommunication company, has reported developing a chipless RFID system that provides a reliable authentic approach for valuable documents, such as intelligence agency reports, financial securities, and banknotes. The system uses "nanometric" materials, tiny particles of chemicals with varying degrees of magnetism, that resonate when bombarded with electromagnetic waves from a reader. Each chemical emits its own distinct radio frequency, or "note," that is picked up by the reader, and all the notes emitted by a specific mix of different chemicals are then interpreted as a binary number. Since the system uses up to 70 different chemicals, each chemical is assigned its own position in a 70-digit binary number [37]. No further information about the working prototype and its applicable tag cost has been released since it was introduced in 2004.

In 2007, Somark Innovation declared that it has a special readable ink as a chipless RFID tag that can be used to track animals, for example, cattle, or even human beings. No further discussion is disclosed by the company and no information is also available on its encoding capacity or the ink (tag) cost [38,39].

A fair comparison among different chipless RFID system techniques with or without metamaterials are provided by Mandel *et al.* [41]. In this communication, various approaches were studied in terms of their merits, frequency band usage, and tag size for one bit of content; then they concluded that the metamaterials can shrink the tag size hence enhancing the merits of the structure. Although in their study the time- and frequency-based systems were considered with the application of metamaterials, no information about the tag cost utilizing metamaterials was provided.

1.5.3 Phase-Domain-Based and Hybrid Chipless RFID Systems

Phase-domain-based system is proposed as a method of data encoding to rely on the phase information of the backscattered signal. In one proposal, the tag consists of some plurality of antenna elements with different dimensions and orientations that provide polarization and phase information [48]. The interrogator device scans the area and uses radar imaging techniques to create an image of the scanned scene. The reradiated RF signals of the tag preferably include polarization and phase information of each antenna element. The proposed tag shapes are presented in Figure 1.7 with dipoles of different length and width. Four different orientations are suggested for each antenna size at angles $0°$, $\pm 45°$, and $90°$, which control their particular phase and polarization responses. It has been also proposed to use some reflective rectangles with two (or more) reactive stub loading that are extended from rectangle antennas. One is perpendicular to the antenna and the other is parallel to the antenna polarization. In this case, the length and position of the reactive stub elements control the phase and polarization parameters of the reflected signal, respectively. It is claimed that the proposed system has the ability to read and track chipless tags 100 m away

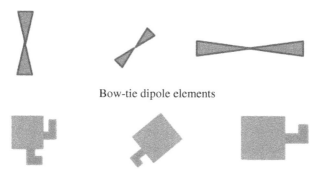

Bow-tie dipole elements

Rectangular reactively loaded patch

Figure 1.7 Phase-domain-based chipless RFID tags.

and to read thousands of tags per second [1,48]. However, no commercial product is available on the market based on this approach yet. This is probably because the content encoding in RFID systems through phase and polarization is very expensive and sometimes impractical. The phase of the reflection by a given element is dependent on its distance to the reader. This means that a slight bending of the tag surface causes a significant phase shift specifically in higher frequencies. There are suggestions to mitigate the phase ambiguity; however, the solutions only apply to chipped tags for low frequencies, below 1 GHz, and no work is reported on chipless systems [49,50].

As a hybrid-based system, the combination of phase deviation and frequency position of the resonances was proposed as the data encoding approach in a chipless RFID system [19]. The tag consists of multiple "C"-shaped resonators that are interrogated through a wideband signal in the band 2.5–7.5 GHz. The claimed content encoding capacity is 23 bits with five "C"-shape resonators. The encoding is based on a matrix of frequency and phase shift deviation. However, the tag was tested in the anechoic chamber and no further discussion was communicated to explore the effect of multipath interference on the system performance. This would be an important aspect of the proposed approach as the reflection from other objects will cause significant effects on the phase shift. There is a need for premeasurement of the scene to calibrate the system as well. This means that the system is vulnerable to errors if any changes happen in the surrounding area of the tag. Moreover, the performance of the tag is not clear regarding to the bending effect. The performance of a real printed tag structure was also not discussed.

In another work, the amplitude- and frequency-based encoding scheme is proposed as a new encoding approach for enhancing the content capacity of the chipless RFID systems [51]. The SRR is the proposed resonator type, and based on the received signal level and the position of the resonances, the data can be encoded. Although the system content capacity is increased, the system complexity and also its sensitive performance toward noise and multipath interference may prevent its actual usage.

1.6   SPATIAL-BASED CHIPLESS RFID SYSTEM

A quick review of the identification technologies, mainly barcodes, QR codes, and RFID systems (chipped and chipless), shows that the market is still waiting for a product meeting all requirements of the industry. While the optical systems, barcodes, and QR codes are suffering from many technological limitations, the chipped RFID systems are not able to reduce their tag price to an acceptable level by industries for main applications. Chipless RFID systems have also limitations on providing a reliable approach for data encoding based on a real printed tag structure with subcent expense. In this situation, any suggestion that fulfils all or main parts of expectations would be welcomed by industries. The new technique of spatial-based system has potentials and abilities that may attract many involved parties in identification field. Moreover, the proposed technique may be combined with other conventional techniques in chipless RFID systems to complement them for a more practical system.

In the spatial-based technique, the tag surface is scanned very precisely by the reader antenna to provide a fine image resolution. Hence, each small section of the tag surface may independently encode the data. The tag is comprised of tiny conductive printable strip or meander lines as EM polarizers. The EM polarizers on the tag surface create a high crosspolar radar cross section (RCS) on the backscattered signal while being interrogated by a linearly polarized signal. The reader utilizes two orthogonally oriented arrays of double-side printed dipole (DSPD) antennas [52,53]. The reader may move around the tag to create a synthetic aperture and provide a fine EM image of the tag using synthetic aperture radar (SAR) signal processing. Alternatively, a multiple input and multiple output (MIMO)-based system may be used for the same effect as SAR technique.

In comparison to other techniques in chipless RFID systems, the tag structure is fully passive and printable. The expected tag printing cost is even less than that of barcodes as the tag size in the proposed technique is smaller than the barcodes. The system does not require any calibration or reference tag. Changing the background of the environment around the tag has no effect on the performance of the proposed system. The tag also provides higher data encoding capacity in a smaller tag size as reported to date by other chipless RFID systems [19,27,46]. The system is very robust to multipath and clutter interference as it is a crosspolar-based system. Moreover, the highly reflective items tagged with the proposed tag structure show a very satisfactory result. All printing inaccuracies are also considered in the proposed system and appropriate solutions are suggested.

1.7   BOOK OUTLINE

The book is divided into two main parts. Part I labeled as "EM Image-Based Chipless RFID System" introduces the novel EM imaging concept for data extraction from chipless RFID tag. The EM-imaging technique exploits advantages of flexible non-line-of-sight (NLoS) operation, high data capacity, low-cost advantages, and fully printable features of the barcodes on low-grade packaging materials. Part II entitled "Advanced Tag Detection Techniques for Chipless RFID Systems" presents smart tag detection techniques for existing chipless RFID systems and an innovative MIMO-based tag detection technique for high content capacity and zero guard-band tag detection. These approaches have been fully developed and tested in Monash Microwave, Antenna RFID, and Sensor Research Group (M.M.A.R.S.) at Monash University.

1.7.1   Part I, EM Image-Based Chipless RFID System

In Part I, the fundamental of EM imaging at the millimeter-wave band for data extraction is introduced followed by the SAR technique. It is shown that the millimeter-wave EM imaging has significant potentials for commercialization of chipless RFID. In this pursuit, the system elements and technical requirements are discussed in detail. The proposed approach to the SAR-based EM-imaging technique enhances the content capacity of the chipless systems to a commercial level, for example, EPC Global Class 1 Generation 2 with 64 data bits. A credit card size EM image-based chipless RFID can encode more than 90 data bits at 60 GHz frequency band. Next, the limitation of the conventional SAR is discussed with MIMO system as solution for addressing the drawbacks of the system. Then the MIMO system is optimized through genetic algorithm approach for minimum hardware complexity. Breakdown of Part I is as follows.

The basics of EM imaging are introduced in Chapter 2. The definitions of range and azimuth resolutions are introduced and their relation with the technical parameter of the system is presented. It is shown that for RFID application, the range resolution suggests impractical requirements while the cross range or azimuth resolution provides a meaningful technique for data retrieval of a chipless RFID tag.

The miniaturized EM polarizers are introduced in Chapter 3. The theoretical working basis of the polarizers on resonance and diffraction basis are discussed. The result of simulation and measurement for strip-line- and meander-line-based polarizers are presented. Then the polarizers are fabricated through different techniques and they are compared regarding their performance.

The advantages of the crosspolar working basis is discussed in Chapter 4. First, the system model is provided for studying the reliability of the system toward noise, multipath, and other environmental factors. Then the fabricated tag is measured to confirm the theoretical expectation. This chapter shows the attributes of the crosspolar working basis for the RFID application.

The reasons and advantages of using the ISM band 60 GHz are discussed in Chapter 5. It is shown that the suggested band provides an acceptable reading range for a chipless RFID system, lowers the tag cost to below the barcode level, and finally suggests high data encoding capacity tag structure. In the second part of this chapter, the reader antenna is introduced along with its technical requirements. A fully printed reader antenna is shown including its design, measurement, and results.

Chapter 6 introduces the conventional SAR technique for imaging of the tag. Technical parameters of the SAR approach for chipless RFID systems are derived. Then tag samples with different encoded data are imaged through the proposed approach. The effect of the synthetic aperture length on the image quality is also studied. Then high data capacity tag is considered in the imaging process, which confirms the ultimate ability of the proposed technique for data encoding purpose of the chipless RFID tag. The downsides of the SAR technique are presented at the end of this chapter.

Chapter 7 addresses the limitations of the SAR technique for a practical application of the spatial-based chipless RFID system. It suggests the MIMO technique to provide a solution for the physical movement of the reader around the tag. While the proposed MIMO system is able to replace the conventional SAR technique, it still requires noticeable reader antennas. The mathematical model of the MIMO system is then considered and it is shown that the suggested MIMO system is not optimized. The genetic algorithm is then applied to the MIMO system to provide the same effect as that of the SAR approach with minimum number of reader antennas. The final system utilizing optimized MIMO system is tested and compared with SAR technique to prove the applicability of the proposed system with minimum hardware complexity.

1.7.2   Part II, Advanced Tag Detection Techniques for Chipless RFID Systems

Part II focuses on advanced tag detection techniques for chipless RFID systems based on maximum likelihood. Chapter 8 gives a brief introduction to Part II and a review of the tag detection techniques reported. Chapter 9 involves both SISO and MIMO tag design and experimental verifications of the proposed tag detection techniques. Chapters 10–12 discuss the proposed tag detection techniques for chipless RFID systems. Chapters 10 and 12 present tag detection techniques for SISO- and MIMO-based chipless RFID systems, respectively, while Chapter 11 presents computationally feasible tag detection techniques. Finally, Chapter 13 summarizes Part II and shares the potential applications, future directions, and recommendations.

In Chapter 8, an introduction to RFID systems and the research aims are presented. The first part of the chapter focuses on two areas in SISO chipless RFID systems. First, it reviews the available chipless RFID tag types and identifies the potential candidate tag types for further investigation. Next, the existing tag detection techniques for chipless RFID systems, their limitations, and areas for improvements are listed. Last part of the chapter presents the state-of-the-art MIMO-based chipped RFID systems followed by the major challenges of MIMO-based chipless RFID systems.

Chapter 9: This chapter presents the designing of two chipless RFID tag types for experimental verification of the proposed detection techniques. The first type is a circular resonator-based SISO chipless RFID tag. The tag is printed on a paper film using a printer with conductive ink. Its performance is verified using measurement data. Then a novel MIMO-based chipless RFID tag was designed in CST and the results were presented.

In Chapter 10, four likelihood-based detection techniques have been presented and their performances have been verified using CST and MATLAB simulation. A fifth tag detection technique is developed for an existing chipless RFID reader and its performances are verified using empirical measurements. The superior performances of the proposed tag detection techniques are compared with the existing detection techniques. The disadvantages of these detection methods are identified and solutions to them are presented in Chapter 11.

Two computationally feasible tag detection techniques are introduced in Chapter 11. The first detection technique is a suboptimal bit-by-bit detection (serial bit reading) in contrast to detecting all the tag bits once (parallel bit reading). The next detection technique is based on trellis tree and Viterbi decoding. This detection method can be incorporated into the proposed tag detection techniques in Chapters 10 and 12.

In Chapter 12, a MIMO-based chipless RFID system is proposed and a MIMO decomposing technique is used for separating the tag responses in each branch. Next, an ML-based tag detection technique is introduced to detect the tag bits encoded in each branch. Its performances are evaluated using MATLAB simulations and then further verified using measurement data.

Chapter 13 summarizes the content provided in Part II on tag detection techniques. Finally, future directions of the research and potential applications are presented.

REFERENCES

1. P. Harrop and R. Das, "Printed and Chipless RFID Forecasts, Technologies & Players 2009–2029," *IDTechEx, USA*, 2009.

2. BiQRious. (2015). *QR Code Capacity*. Available: http://qrcodetracking.com/qr-code-capacity/.

3. *IDAutomation group*. Available: http://www.idautomation.com/barcoding4beginners.html.

4. C. Turcu, *Current Trends and Challenges in RFID*, InTech, 2011.

5. T. Lotlikar, R. Kankapurkar, A. Parekar, and A. Mohite, "Comparative Study of Barcode, QR-Code and RFID System," *International Journal Computer Technology & Applications*, vol. 4, pp. 817-821, 2013.

6. E. Arendarenko, "A Study of Comparing RFID and 2D Barcode Tag Technologies for Pervasive Mobile Applications," Master, Department of Computer Science and Statistics, University of Joensuu, 2009.

7. G. White, G. Gardiner, G. P. Prabhakar, and A. Razak, "A comparison of barcoding and RFID technologies in practice," *Journal of Information, Information Technology and Organizations*, vol. 2, pp. 119-132, 2007.

8. Cartman. (Feb 5, 2015). *Asset Tracking: RFIDs, Barcodes or QR Codes?* Available: https://www.webcheckout.net/blog/rfid-asset-tracking/.

9. A. Campbell. (2011). *QR Codes, Barcodes and RFID: What's the Difference?* Available: http://smallbiztrends.com/2011/02/qr-codes-barcodes-rfid-difference.html.

10. Cisco. (Oct 14, 2014). *Internet-of-Everything*. Available: http://www.cisco.com/c/r/en/anz/internet-of-everything-ioe/tomorrow-starts-here/index.html.

11. S. Preradovic and N. Karmakar, "Chipless RFID, Bar Code of the Future," *IEEE Microwave Magazine*, vol. 11, pp. 87-97, 2010.

12. R.R. Fletcher. (2002). *Low-Cost Electromagnetic Tagging: Design and Implementation. PhD thesis*. Available: www.media.mit.edu/physics/publications/theses/97.02.fletcher.pdf.

13. S. Preradovic, "Chipless RFID System for Barcode Replacement," PhD, ECSE, Monash University, 2009.

14. V. Chawla and D.S. Ha, "An Overview of Passive RFID," *Communications Magazine, IEEE*, vol. 45, pp. 11-17.

15. P. Harrop. (2006). *The Price-Sensitivity Curve for RFID*. Available: http://www.printedelectronicsworld.com/articles/the-price-sensitivity-curve-for-rfid-00000488.asp?sessionid=1.

16. C. S. Hartmann, "A global SAW ID tag with large data capacity," in *IEEE Ultrasonic Symposium*, 2002.

17. S. Preradovic, I. Balbin, N.C. Karmakar, and G.F. Swiegers, "Multiresonator-Based Chipless RFID System for Low-Cost Item Tracking," *IEEE Transactions on Microwave Theory and Techniques*, vol. 57, pp. 1411–1419, 2009.

18. T. Singh, S. Tedjini, E. Perret, and A. Vena, "A frequency signature based method for the RF identification of letters," in *2011 IEEE International Conference on RFID*, 2011, pp. 1–5.

19. A. Vena, E. Perret, and S. Tedjini, "Chipless RFID Tag Using Hybrid Coding Technique," *IEEE Transactions on Microwave Theory and Techniques*, vol. 59, pp. 3356–3364, 2011.

20. M.S. Bhuiyan, R.E. Azim, and N. Karmakar, "A novel frequency reused based ID generation circuit for chipless RFID applications," in *Asia-Pacific Microwave Conference (APMC)*, Melbourne, Australia, 2011, pp. 1470–1473.

21. I. Balbin and N.C. Karmakar, "Phase-Encoded Chipless RFID Transponder for Large-Scale Low-Cost Applications," *IEEE Microwave and Wireless Components Letters*, vol. 19, pp. 509-511, 2009.

22. Md. Aminul Islam and N.C. Karmakar, "A Novel Compact Printable Dual-Polarized Chipless RFID System," *IEEE Transactions on Microwave Theory and Techniques*, vol. 60, pp. 2142-2151, 2012.

23. A. Ramos, A. Lazaro, D. Girbau, and R. Villarino, "Time-Domain Measurement of Time-Coded UWB Chipless RFID Tags," *Progress in Electromagnetics Research*, vol. 116, pp. 313–331, 2011.

24. Z. Lu, S. Rodriguez, H. Tenhunen, and Z. Li-Rong, "An innovative fully printable RFID technology based on high speed time-domain reflections," in *High Density Microsystem Design and Packaging and Component Failure Analysis*, 2006, pp. 166–170.

25. V.P. Plessky and L.M. Reindl, "Review on SAW RFID Tags," *IEEE Transactions on Ultrasonics, Ferroelectrics, and Frequency Control*, vol. 57, pp. 14-23, 2010.

26. C. Mandel, M. Schüßler, M. Maasch, and R. Jakoby, "A novel passive phase modulator based on LH delay lines for chipless microwave RFID applications," in *IEEE MTT-S International Microwave Workshop on Wireless Sensing, Local Positioning, and RFID*, Croatia, 2009.

27. S. Gupta, B. Nikfal, and C. Caloz, "Chipless RFID System Based on Group Delay Engineered Dispersive Delay Structures," *IEEE Antenna and Wireless Propagation letters*, vol. 10, pp. 1366-1368, 2011.

28. F.J.H. Martínez, F. Paredes, and G. Zamora, "Printed Magnetoinductive-Wave (MIW) Delay Lines for Chipless RFID Applications," *IEEE Transactions on Antennas and Propagation*, vol. 60, pp. 5075–5082, 2012.

29. A. Chamarti and K. Varahramyan, "Transmission Delay Line Based ID Generation Circuit for RFID Applications," *IEEE Microwave and Wireless Components Letters*, vol. 16, pp. 588-590, 2006.

30. M. Schüußler, C. Damm, M. Maasch, and R. Jakoby, "Performance Evaluation of Left-Handed Delay Lines for RFID Backscatter Applications," in *Microwave Symposium Digest*, 2008.

31. J. Vemagiri, A. Chamarti, M. Agarwal, and K. Varahramyan, "Transmission Line Delay-Based Radio Frequency Identification (RFID) Tag," *Microwave and Optical Technology Letters*, vol. 49, p. 1900, 2007.

32. RF SAW, Inc. (14 July 2014). *RF SAw Inc.* Available: http://www.rfsaw.com.

33. S. Preradovic, I. Balbin, N.C. Karmakar, and G. Swiegers, "Chipless Frequency Signature Based RFID Transponders," in *1st European Wireless Technology Conference*, Amsterdam, 2008, pp. 302-305.

34. D. Girbau, J. Lorenzo, A. Lázaro, C. Ferrater, and R. Villarino, "Frequency-Coded Chipless RFID Tag Based on Dual-Band Resonators," *IEEE Antennas and Wireless Propagation Letters*, vol. 11, pp. 126-128, 2012.

35. F. Costa, S. Genovesi, and A. Monorchio, "A Chipless RFID Based on Multiresonant High-Impedance Surfaces," *IEEE Transactions on Microwave Theory and Techniques*, vol. 61, 2013.

36. S. Mukherjee and G. Chakraborty, "Chipless RFID Using Stacked Multilayer Patches," in *Applied Electromagnetics Conference (AEMC), 2009, Kolkata*, 2009.

37. M. Glickstein, "Firewall Protection for Paper Documents," *RFID Journal Internet Article*, 2004.

38. K.C. Jones, "Invisible RFID Ink Safe For Cattle And People, Company Says," *InformationWeek*, 2007.

39. St. Louis, "SOMARK's Chipless RFID Ink Tattoo Field Demo Brings the Company Closer to Launch," 2008.

40. C. Mandel, B. Kubina, M. Schussler, and R. Jakoby, "Group-Delay Modulation with Metamaterial-Inspired Coding Particles for Passive Chipless RFID," presented at the *RFID-TA*, Nice, 2012.

41. C. Mandel, B. Kubina, M. Schüßler, and R. Jakoby, "Metamaterial-inspired Passive Chipless Radio-Frequency Identification and Wireless Sensing," *Annals of Telecommunications*, vol. 68, pp. 385-399, 2013.

42. S. Kim, A. Georgiadis, A. Collado, and M.M. Tentzeris, "An Inkjet-Printed Solar-Powered Wireless Beacon on Paper for Identification and Wireless Power Transmission Applications," *IEEE Transactions on Microwave Theory and Techniques*, vol. 60, 4178–4186.

43. I. Jalaly and I.D. Robertson, "RF Barcodes Using Multiple Frequency Bands," in *International Microwave Symposium Digest, IEEE MTT-S*, 2005.

44. I. Balbin, "Chipless RFID Transponder Design," PhD, Monash University, 2010.

45. H.S. Jang, W.G. Lim, K.S. Oh, S.M. Moon, and J.W. Yu, "Design of Low-Cost Chipless System Using Printable Chipless Tag With Electromagnetic Code," *IEEE Microwave and Wireless Components Letters*, vol. 20, 2010.

46. R. Rezaiesarlak and M. Manteghi, "Complex-Natural-Resonance-Based Design of Chipless RFID Tag for High-Density Data," *IEEE Transactions on Antenna and Propagation*, vol. 62, 2014.

47. B.A. Munk, *Frequency Selective Surfaces, Theory and Design*: John Wiley & Sons, Inc., 2000.

48. M.G. Pettus, "RFID System Utilizing Parametric Reflective Technology," USA Patent, 2005.

49. S. Särkkä, V.E.V. Viikari, M. Huusko, and K. Jaakkola, "Phase-Based UHF RFID Tracking With Nonlinear Kalman Filtering and Smoothing," *IEEE Sensors Journal*, vol. 12, 904–910, 2012.

50. P.V. Nikitin, R. Martinez, S. Ramamurthy, H. Leland, G. Spiess, and K.V.S. Rao, "Phase Based Spatial Identification of UHF RFID Tags," presented at the *IEEE RFID*, 2010.

51. Md. S. Bhuiyan, A. Azad, and N. Karmakar, "Dual-band Modified Complementary Split Ring Resonator (MCSRR) Based Multi-resonator Circuit for Chipless RFID Tag," in *IEEE – Intelligent Sensors, Sensor Networks and Information Processing*, Melbourne, Australia, 2013.

52. M. Zomorrodi and N.C. Karmakar, "An Array of Printed Dipoles at 60 GHz," in *IEEE International Symposium on Antennas and Propagation*, Memphis, Tennessee, USA, July 2014, pp. 73–74.

53. M. Zomorrodi and N. Karmakar, "A Low Cost Wideband Printed Dipole Array Antenna for 60 GHz Image-Based Chipless RFID Reader," *IEEE Antennas and Propagation Magazine*, vol. 57, p. 13, 2015.

2

EM IMAGING

This chapter reviews the fundamentals of electromagnetic (EM) imaging and considers the possibility of EM-imaging technique as data encoding approach in chipless radiofrequency identification (RFID) systems. The resolution of EM image is fully investigated and related to the technical parameters of the imaging device with a specific discussion for tag applications. The synthetic aperture radar (SAR) technique is then introduced as an almost mandatory approach to many EM-imaging applications. The usage of EM imaging in different applications is briefly introduced with a short description of the technical characteristics of each. Finally, it concludes that the EM-imaging technique is technically applicable for content coding in a chipless RFID system.

2.1 EM-IMAGING FUNDAMENTALS

The ultimate goal of imaging is to gain information about an object through its image. Basic radar system and optical imaging device may define the two extents of imaging. With a radar, the main emphasis is to detect the presence of the target based on its reflected energy. Hence, a point on the radar display that is corresponding to a specific target may be interpreted as the target image though it provides minimum information

Advanced Chipless RFID: MIMO-Based Imaging at 60 GHz–ML Detection, First Edition.
Nemai Chandra Karmakar, Mohammad Zomorrodi, and Chamath Divarathne.
© 2016 John Wiley & Sons, Inc. Published 2016 by John Wiley & Sons, Inc.

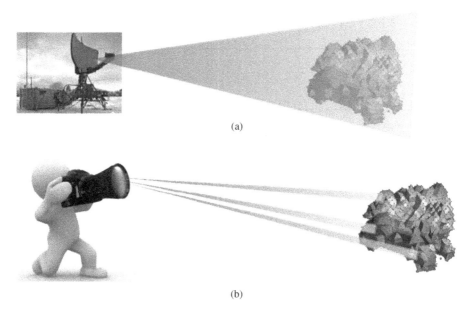

(a)

(b)

Figure 2.1 The precision comparison of two imaging systems: (a) conventional radar and (b) optical camera.

about the target. On the other hand, a camera as an optical imaging system provides much more detailed information about the target than the radar system. Figure 2.1 figuratively compares a radar system and a camera as more advanced imaging system. Therefore, it may be said that the imaging is characterization of a physical object based on the analysis of the return signal. Obviously, the ability of an imaging system to distinguish the reflections from multiple points of a physical object, known as the spatial resolution of the system, governs the quality of the imaging system.

To evaluate the applicability of imaging techniques for chipless RFID systems, it is mandatory to quantify the required spatial resolution for the data encoding purpose. The required image quality or the accuracy of scanning directly relates to the associated data encoding capacity in the chipless RFID tags. Therefore, to find out what is the minimum required scanning precession, one should know the expected content capacity of the chipless RFID system. On the other hand, the size of the tag is limited to certain values. For many RFID applications, the size of a credit card ($85 \times 54\,\text{mm}^2$) appears to be the maximum allowable size of the RFID tag [1,2].

The expected content capacity of chipless RFID systems is more difficult to standardize as every industry has its own requirements and expectations. However, many research entities assume 50-bit data encoding capacity for a chipless tag structure as the ultimate capacity of these systems [3–5]. This may suggest 1 bit/cm^2 would be a very satisfying data encoding capacity for a chipless RFID system.

$$\text{Encoding capacity} = \frac{50}{8.5 \times 5.4} \approx 1\,\text{bit/cm}^2 \qquad (2.1)$$

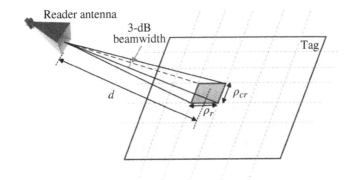

Figure 2.2   Reader footprint on tag structure.

Therefore, the targeted scanning precession should be around 10 mm. This means that the reflected signal from each 1 cm^2 of the tag surface shall be processed separately by the reader. The easiest approach doing this is to utilize a reader antenna in which its radiation pattern illuminates every 10 mm of the tag surface separately. This illumination area is known as the *footprint* of the reader on the tag surface as shown in Figure 2.2. Two resolutions can be associated with the reader footprint: range and cross-range resolutions. These two are discussed in the following sections.

2.2   RANGE RESOLUTION

The reader's ability to separate two targets in range is known as *range resolution*. Based on the signal type used in the transmit section of the radar, the range resolution may vary. For a single short pulse, the reader distinguishes two adjacent targets in range, if the separation time between their echoes is more than the pulse width. The range resolution, ρ_r, is the minimum distance between two adjacent targets separated by the reader and is defined by

$$\rho_r = \frac{cT_p}{2} \tag{2.2}$$

where c is the speed of light and T_p is the pulse width. Figure 2.3 shows the range resolution in a more understandable configuration. Expression (2.2) is only valid if the

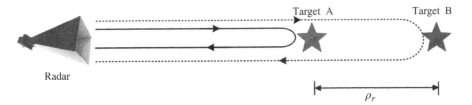

Figure 2.3   Range resolution concept.

targets and the radar system are in line as shown in Figure 2.3. With a fine range resolution, shorter pulses are needed. However, to maintain a satisfactory average system
transmitted power, these shorter pulses shall acquire higher power level; otherwise,
the system signal-to-noise ratio (SNR) will drop.

It is very difficult to generate high-power narrow pulses specifically in higher
frequency ranges. This means that narrower pulses, which are required for finer
range resolutions, contradict the requirement of a higher average transmitted
signal for maintaining the minimum SNR for a fixed reading range. To satisfy
both requirements, it is possible to have long, phase-coded pulses, which can be
compressed with a signal processing operation after reception. In this scenario, a
linear frequency-modulated (FM) signal, known as chirp signal, is transmitted for
a long enough time as a pulse signal to satisfy the minimum transmission power
requirement. However, signal processing after reception provides a fine range
resolution corresponding to a very narrow pulse shape.

This chirp signal is a common choice for a transmitted signal by many modern
radar systems. In the chirp signal approach, the linear FM waveform exists over
a pulse length of T_p that is adequately long for the minimum transmission power
requirement. In this time frame, the frequency of the transmit signal, which varies
with time, is a function of the chirp rate, γ. The instantaneous frequency is then

$$f_c + \gamma(t - nT) \tag{2.3}$$

where f_c is the center frequency and $t = nT$ signifies the center of the pulse. The
related bandwidth B of this signal is γT_p. From the signal theory, a pulse with bandwidth B corresponds to the time duration of $1/B$. Hence, from the above expression,
the new range resolution of the chirp-based radar is

$$\rho_r = \frac{c}{2B} \tag{2.4}$$

This means that the range resolution of the system is enhanced by factor B after
signal processing. This range resolution is much better than the initial range resolution
corresponding to the physical length of the transmitted pulse, T_p.

The range compression factor is the ratio of the pulse's length before time compression (T_P) to its length after compression $(1/B)$. The normal value of pulse compression ratio (PCR) is in order of $1e^5$. As an advantage of the linear FM signal, the
analog bandwidth of the received signal is decreased; hence, the required speed of an
analog-to-digital (A/D) convertor unit is also slowed:

$$PCR = T_p\, B = \gamma T_p^2 \tag{2.5}$$

The assumption of having the radar system and the targets in one line is not always
practically valid. In fact, in many applications, the radar has a slant view angle to the
target, for example, earth imaging. In this case, Equations (2.4) and (2.5) equivalently
are not showing the final range resolution of the system. Instead, the final range resolution known as the ground-range resolution could be easily found from its equivalent

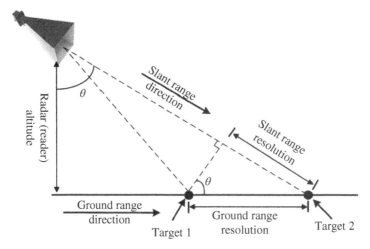

Figure 2.4 Slant- and ground-range resolutions.

slant range resolution through

$$\rho_{gr} = \frac{\rho_r}{\sin \theta} \tag{2.6}$$

where ρ_{gr} is the ground range resolution that is the final resolution of the radar system on the target scene and ρ_r is the slant range resolution. Figure 2.4 clearly shows these two range resolutions. For the RFID application, the same scenario is valid. The angle between the reader antenna and the tag's surface is almost 90° as shown in Figure 2.5, hence the final range resolution shall be based on Equation (2.6). Considering the normal reading range of chipless RFID systems, achieving the ground range resolution of few millimeters on the tag surface requires huge bandwidths in tens or even hundreds of gigahertz that is technically impractical to achieve. Therefore, it can be concluded that providing the fine range resolution of few millimeters on the tag surface is not possible due to the limited available frequency band of operation. Hence, the range resolution has no benefit for data encoding purposes on the chipless RFID tag.

2.3 CROSS-RANGE OR AZIMUTH RESOLUTION

Two targets in cross-range or azimuth can be separated by the reader if their angular separation is greater than the reader beamwidth as shown in Figure 2.2. This is known as the reader cross-range or azimuth resolution. This means that reader is not able to resolve targets that are closer to each other than the reader beamwidth. The azimuth resolution simply corresponds to d and 3-dB beamwidth ($\theta_{3\,\mathrm{dB}}$) of the antenna through

$$\rho_A = d \cdot \tan(\theta_{3\,\mathrm{dB}}) \tag{2.7}$$

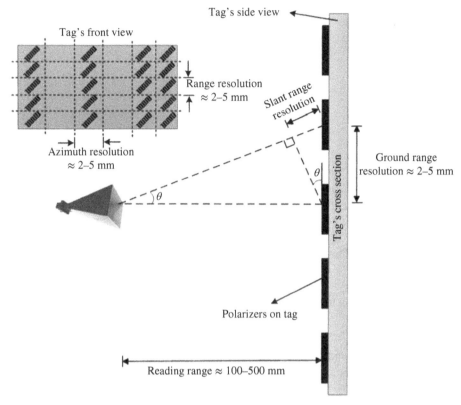

Figure 2.5 Ground- and slant-range resolutions of the millimeter-wave chipless tag.

with d as reading range and $\theta_{3\,dB}$ showing the antenna beamwidth, the ρ_A is the achieved azimuth resolution. The required antenna beamwidth based on the normal reading distance of chipless RFID systems is depicted in Figure 2.6, which shows that 1–6° 3 dB beamwidth is necessary.

Creation of such a narrow beamwidth requires a very large antenna size as the beamwidth inversely relates to the antenna physical size. The exact relation between the antenna's beamwidth and its physical size depends on the aperture illumination function, which controls the main beam and side lobes of the antenna. However, a practical rule of thumb for typical illumination functions expresses the beamwidth in terms of aperture size and the wavelength as follows:

$$\theta_{3\,dB} = \alpha \frac{\lambda}{D} \tag{2.8}$$

where θ is the 3-dB beamwidth of the antenna in radians, λ is the wavelength, D is the antenna physical size, and α is a constant reflecting the main lobe widening due to the weighting aperture. For a uniform aperture weighting, $\alpha \approx 0.89$ and hence (2.8) is changed to $\theta_{3\,dB} = 50(\lambda/D)$ when θ is measured in degrees. By assuming

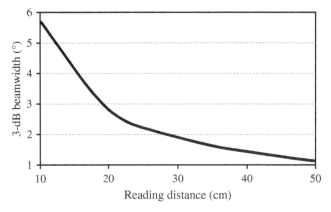

Figure 2.6   Required 3-dB beamwidth of reader antenna versus reading distance.

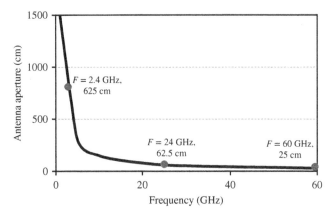

Figure 2.7   Antenna aperture size for 1° beamwidth.

1° beamwidth demand, the required aperture size versus frequency is depicted in Figure 2.7. The actual required antenna size is also shown for three common ISM bands of 2.4, 24, and 60 GHz.

Among the three available ISM bands, 2.4, 24, and 60 GHz, only the 60 GHz band offers an antenna size that may be practical for a reading distance of 10–50 cm. In frequencies below 60 GHz, the required aperture size of the antenna is much bigger than the reading distance, hence are nonpractical scenario for the tag imaging purpose. This is the main reason that the millimeter-wave band of 60 GHz is selected for the proposed image-based chipless RFID system.

2.4   SYNTHETIC APERTURE RADAR (SAR) NECESSITY

As discussed in the previous section, achieving 1–6° antenna beamwidth at millimeter-band 60 GHz mandates the antenna aperture size of only 25 cm that seems to be practical. Therefore, it appears that a single antenna would be sufficient for the required precision scanning of the tag. However, one should note that the calculated beamwidth by Equation (2.8) is only valid for the far field region of the antenna. If the measurement point locates inside the near field, then it is unlikely to expect such a narrow beamwidth. The far field distance of an antenna with an aperture size of 25 cm at 60 GHz is 25 m that is far beyond the reading zone of the chipless RFID systems:

$$R_{\text{far field}} = \frac{2D^2}{\lambda} = \frac{2 \times (0.25)^2}{5 \times 10^{-3}} = 25\,\text{m} \tag{2.9}$$

Therefore, it appears that the suggestion for a fine image resolution at azimuth is also failed to provide a practical approach for chipless RFID systems.

EM imaging of earth by a radar system that is normally mounted on an air-/spacecraft has the same limitation. A super narrow beam antenna is required at the craft station to precisely scan the earth surface. However, based on the antenna theory, this super narrow beamwidth radiation mandates a massive antenna with few-kilometer size. Fabrication of such antennas with kilometer size is extremely difficult and handling it by an air-/spacecraft station is impractical. Instead of utilizing an impractical big antenna, Wiley [6] proposed to simply move a small antenna around the earth and constructively add the received signal at different positions of the radar system. This concept is called synthetic aperture radar (SAR) to signify that a signal synthesis accomplishes what would otherwise require a large antenna aperture. In other words, a large synthetic aperture antenna is formed by a moving small antenna over the target region. The target is repeatedly illuminated with radio wave pulses at different frequencies, and the echo waveforms received successively at the different antenna positions are coherently detected and stored, and then postprocessed together to resolve elements in an image of the target region. The obtained EM image of the target scene through SAR technique acquires a much finer spatial resolution than is possible with conventional beam-scanning means.

To simplify the idea of SAR technique for achieving fine scanning resolutions, one may refer to Figure 2.8 that shows the transition from a conventional radar system to an SAR-based system. Figure 2.8(a) presents a conventional radar antenna with large enough aperture size providing the demanded scanning resolution. This antenna can be replaced by an array antenna with large number of elements. The higher number of elements, the more similarity between two antennas in Figure 2.8(a) and (b). However, thousands of elements may be required to provide an equivalent array antenna for the physical antenna. Instead of utilizing a huge number of elements, SAR technique suggests to use only one element and sequentially move it to different positions in the array configuration. By restoring the received signal in each individual position, the effect of array antenna with thousands of elements, Figure 2.8(b), or equivalently the effect of the real antenna in Figure 2.8(a) can be successfully constructed.

For the application of tag imaging, the size of antenna, 25 cm, is easily manageable; however, its long far field mandates the use of the SAR technique. The received

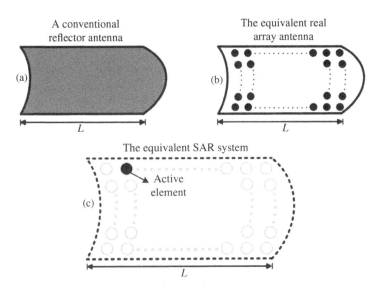

Figure 2.8 Evolution of conventional radar toward SAR system through array antenna concept.

signal at each individual position of the real antenna is added coherently with proper phase adjustment to maintain the large synthetic aperture, which yields a narrow beamwidth after proper signal processing. This means that physically scanning the target scene through a small antenna with a wide radiation pattern is equivalent to scanning the target scene with a much narrower radiation pattern in the processor.

The signal processing technique that SAR-based systems use to achieve a fine azimuth resolution is the feature that distinguishes the SAR-based systems from other imaging techniques [144, 147]. Synthetic aperture processing in azimuth exploits the linear FM modulation that the quadratic variation in range to a target introduces as the radar passes. The processor is able to compress this modulated azimuth signal to generate the resolution associated with a narrow beamwidth from a wide beamwidth antenna [144].

For the chipless RFID application, the relation between the required synthetic aperture length and achievable azimuth resolution based on various reading ranges, R, is shown in Figure 2.9. As one can easily see, the necessary length of the synthetic aperture is completely practical to achieve the azimuth resolution of few millimeters. For example, in 15 cm reading distance, a 20 cm synthetic aperture is enough to provide less than 4 mm azimuth resolution on the tag surface. A 25 cm synthetic aperture length is normally enough to provide the azimuth resolution of less than 5 mm for a reading range up to 25 cm. As can be seen later in final stage of the proposed technique, the implementation of the system and the achieved image resolution confirm the values of the graphs in Figure 2.9. This means that although the range resolution of few millimeters requires a huge spectrum bandwidth, the azimuth resolution is quite practical and proposes a new method for data encoding on a chipless tag with noticeable data capacity.

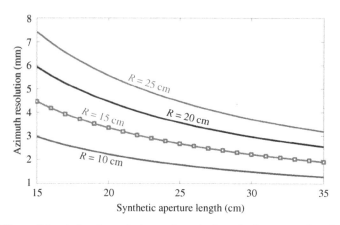

Figure 2.9 Azimuth resolution and required synthetic aperture length.

For applications such as earth imaging, the SAR technique is the only solution as carrying a large antenna over the earth is not practical. There are, however, other applications on which the use of large antennas is feasible for imaging purposes, but the SAR technique is preferred for different reasons. For instance, in the body imaging application, it is possible to use a large antenna providing a fine image resolution of the body; however, the SAR technique is much preferred as it minimizes the amount of radiation on the body. Most body imaging systems nowadays use the SAR-based technique to comply with EM-radiation regulations [75–77]. Through-to-wall imaging (TWI) based on the SAR technique is also extensively used in construction activities [78] as well as intelligence and security applications.

2.5   EM IMAGING FOR CONTENT CODING

As has been already discussed, the main goal of EM imaging for chipless RFID tag data encoding is to provide as fine image resolution as possible. It is also shown that each square centimeter of the tag surface shall provide at least one bit of data to address the content encoding capacity expectation of a chipless RFID in a credit card size. Therefore, the tag surface may be divided into a number of pixels, each pixel with 1 cm^2 size. The ultimate aim of the reader is to illuminate each pixel individually. Figure 2.10 shows the pixelated tag surface. The width of the tag has m pixels and its length is divided into n pixels. Without losing generality, it is assumed that the pixels are square with 1 cm length. This means that the tag surface comprises a matrix of $m \times n$ pixels. As each pixel carries 1 bit of data, the total data encoding capacity would be up to mn^2 if the reader is able to precisely scan each pixel individually as shown in Figure 2.10 with the right-hand antenna.

It was already shown that while precise image resolution in the azimuth direction is completely practical through SAR technique, the same resolution in the range direction requires impractical spectrum bandwidth. It is claimed that the range resolution

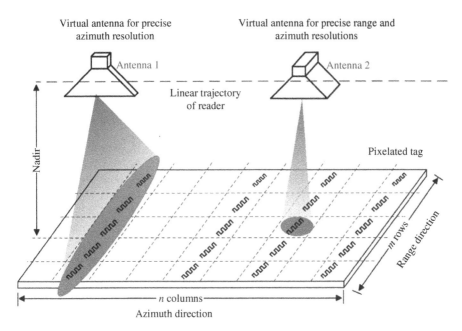

Figure 2.10 Pixelated tag and two different reading scenarios.

of the reader can be significantly improved without enormous bandwidth requirement [149]. As no technical details of the referred work have been revealed yet, therefore at this stage, it can be concluded that providing the millimeter range resolution is not feasible for the proposed application with a reasonable amount of bandwidth.

Ignoring the fine range resolution means that the detectable pixels are rectangles rather than squares. Each pixel has a millimeter width, in the azimuth direction, and length in order of tens of centimeters, in the range direction. This means that the presence or absence of a group of polarizers on each strip of the tag represents 1/0 encoded data. Hence, the content capacity of the tag drops to 2^n and m has no effect on the encoding capacity. The higher the number of polarizers on each column of the tag surface, the higher the reflected signal level is. As every column of the polarizers on the tag surface represents 1 bit of data, then the content capacity is proportional to the tag length. This scenario that is practical is shown with the left-hand antenna in Figure 2.10.

2.6 CONCLUSIONS

In this chapter, the fundamentals of the EM imaging have been discussed. First, the image resolutions, range, and azimuth were presented and their relation with technical parameters of the systems was shown. Then the expected image resolution for RFID application was calculated. Based on the required image resolution, it was found that the range resolution results in impractical system specification. However, the azimuth resolution of millimeter order is possible at 60 GHz and higher frequency bands. Moreover, it was found that one single antenna is not practical to provide the expected azimuth resolution on the tag surface. Instead, a small antenna shall physically move around the tag and create a large synthetic aperture to provide the required azimuth resolution. Based on the findings in this chapter, the final data encoding capacity of the proposed system has been determined.

REFERENCES

1. S. Preradovic and N. Karmakar, "Chipless RFID, bar code of the future," *IEEE Microwave Magazine*, vol. 11, pp. 87-97, 2010.

2. Md. Aminul Islam and N.C. Karmakar, "A Novel Compact Printable Dual-Polarized Chipless RFID System," *IEEE Transactions on Microwave Theory and Techniques*, vol. 60, pp. 2142–2151, 2012.

3. S. Preradovic, I. Balbin, N.C. Karmakar, and G.F. Swiegers, "Multiresonator-Based Chipless RFID System for Low-Cost Item Tracking," *IEEE Transactions on Microwave Theory and Techniques*, vol. 57, pp. 1411-1419, 2009.

4. T. Singh, S. Tedjini, E. Perret, and A. Vena, "A Frequency Signature Based Method for the RF Identification of Letters," in *2011 IEEE International Conference on RFID*, 2011, pp. 1–5.

5. A.T. Blischack and M. Manteghi, "Embedded Singularity Chipless RFID Tags," *IEEE Transactions on Antenna And Propagation*, vol. 59, pp. 3961–3968, 2011.

6. C.A. Wiley, "Synthetic Aperture Radars: A Paradigm for Technology Evolution," *IEEE Transactions on Aerospace and Electronic Systems*, vol. AES-21, p. 4, 1985.

3

TINY POLARIZERS, SECRET OF THE NEW TECHNIQUE

3.1 INTRODUCTION

The idea of electromagnetic (EM) imaging for data encoding in chipless radiofre-quency identification (RFID) systems was briefly introduced in Chapter 2. It was shown that imaging technique differs from other conventional methods as it provides extra spatial diversity. In imaging approach, each small section of the tag is separately scanned while it independently carries data. Providing fine scanning resolution governs the frequency band of operation and mandates the synthetic aperture radar (SAR) technique. However, the next issue that shall be answered is how to encode data in a small area of millimeter order.

Earth imaging system, the first and most important application of EM imaging, may suggest an approach for data encoding. In earth imaging, the earth's surface is interrogated by a radar system that is normally mounted on an air-/spacecraft. Based on the reflected signal, the radar system forms the related EM image of the earth. The strength of the reflected signal from each section of the earth's surface is linked to the image and shows the terrain of the earth. Rough surfaces result in higher reflection and therefore a brighter part in the EM image, for example, a building or mountain. An impinged signal on a flat surface is reflected in the opposite direction, by Snell's law; hence, radar receives no energy and a dark section in the image is formed. Figure 3.1 shows the principle of image brightness in an earth imaging system. This may initiate the idea of shaping the RFID tag surface in such a way to create a unique image when illuminated by a proper EM signal.

Advanced Chipless RFID: MIMO-Based Imaging at 60 GHz–ML Detection, First Edition.
Nemai Chandra Karmakar, Mohammad Zomorrodi, and Chamath Divarathne.
© 2016 John Wiley & Sons, Inc. Published 2016 by John Wiley & Sons, Inc.

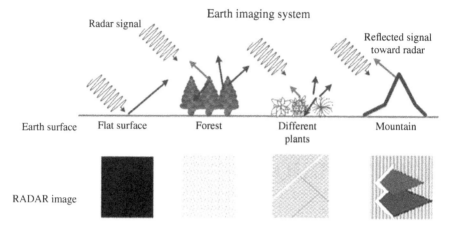

Figure 3.1 Reflection strength based on terrain roughness in earth imaging system [1].

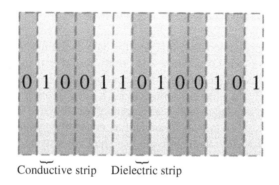

Figure 3.2 Initial proposed tag structure based on earth imaging concept.

Considering the tag cost expectation of less than one cent (1¢), the tag shall be shaped by only combining conductive and nonconductive sections. Moreover, elements on the tag structure shall be very simple; otherwise, system performance will be significantly degraded due to the noticeable printing errors on the millimeter-band 60 GHz. The simplest idea is a structure similar to a barcode as shown in Figure 3.2. Conductive and nonconductive strips may create different levels of reflection similar to earth imaging process; hence a unique EM image will be resulted.

To examine this idea, two separate conductive and dielectric strips are simulated through 3D full-wave CST software at the band 57–64 GHz. The strips are illuminated by a linearly polarized plane wave and the backscattered signals on the reader side are measured. The results are shown in Figure 3.3 with 30 cm reading distance. The incident electric field is 1 V/m, and it is parallel to the main axis of the strip. Interestingly, reflection from a conductive strip is higher than that of the dielectric strip, which contradicts with what was expected from the earth imaging system. Two

SWEETNESS OF DIFFRACTION   **39**

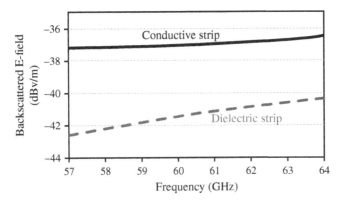

Figure 3.3 Backscattered signals from different types of strips; simulation result.

reasons may be suggested. First, the illumination angle between the reader antenna and the tag is 90°. This means that all of the impinged energy to the conductive strip is reflected back toward the reader according to Snell's law. While for the dielectric surface, some of the incident energy is spread out due to its rougher surface. A squint angle other than 90° results in higher reflection toward the radar from a rougher surface as shown in Figure 3.1 for the earth imaging application. Second, the received signal in the earth imaging is only based on the reflected signal. While for the RFID tag imaging, the impinging signal induces a surface current on the conductive strip surface, which acts as a second source of electromagnetic emission.

Irrespective of higher reflection from a conductive strip than a dielectric, the main issue is the very close signal level of the two strips type. The difference between two scenarios is less than 3 dBm. This is not a reliable field strength difference especially considering that it is a simulation result. It would be very difficult to separate reflections from conductive and dielectric strips in a real scenario, especially when the clutter and multipath effect of the surrounding objects are considered. Changing the polarization of the incident E-field does not affect the ratio of the received signal from conductive and dielectric strips.

While it appears that the signal-level approach similar to the earth imaging application is not able to provide a reliable way for tag imaging, the unique behavior of a conductive strip suggests an interesting approach to data encoding on the tag surface, which is discussed in the following section.

3.2 SWEETNESS OF DIFFRACTION

Finding a reliable approach for tag shaping, strips with varying widths, as shown in Figure 3.2, are interrogated at various illumination angles and polarization schemes. Although the signal levels have not changed enough to provide a reliable approach for the data encoding algorithm, a very interesting phenomenon was observed during the simulation. It was noticed that the backscattered signal from a narrow conductive strip shows elliptical polarization behavior while the strip is illuminated by a linearly polarized signal. This means that the conductive strip is capable of changing the polarization of the incident signal. It was found that for a narrow conductive strip, the backscattered signal may include a polarization component that is orthogonal to the polarization of the incident signal; hence the conductive strip acts as an EM polarizer. The depolarization aspect of the conductive strip suggests working on a cross-polar basis instead of a copolar basis. The theory behind this behavior of the conductive strip as an EM polarizer is discussed.

3.2.1 Diffraction, Kobayashi Potentials

Different phenomena are involved when a conductive strip is illuminated by an EM signal, reflection, refraction and diffraction, for example. As a strip has a finite surface with straight edges, it can behave differently based on the polarization and frequency of the incident wave as well as strip size and its orientation. Reflection based on Snell's law is the normal behavior that is expected for a wide strip, compared to the wavelength. However, when the strip is narrow enough, diffraction would be of more concern and act as the dominant phenomenon. In the past decades, many studies investigated diffraction from a metallic strip or a slot in a metal plane using various numerical and analytical techniques [2–9]. All research and studies on diffraction can be summarized as follows [10]:

- Geometric optics (GO) [3]
- Physical optics (PO) [4]
- Geometrical theory of diffraction (GTD) [5]
- Uniform asymptotic theory of diffraction (UAD) [6]
- Uniform theory of diffraction (UTD) [5]
- Physical theory of diffraction (PTD) [7]
- Spectral theory of diffraction (STD) [8]
- Method of equivalent current (MEC) [9]
- Hybrid methods.

Each method has its own advantages and limitations. Among them, geometrical optics (GO), PO, and uniform theory of diffraction (UTD) are the most common approaches. The Kobayashi potential (KP) method is an analyticonumerical technique used as an alternative to previously mentioned techniques with more accurate results and less procedure complexity [11]. Diffraction by a conductive strip on a dielectric slab has been thoroughly studied by Imran [12]. The far field pattern and induced surface current on an infinite strip etched to a dielectric slab based on the polarization of the incident plane wave is formulated in this work.

Figure 3.4 shows the geometry of the diffraction phenomenon where a conductive strip with negligible thickness etched on a dielectric slab is illuminated with a plane wave. The plane wave incident angle is Φ_0. The strip width is $2a$ and it is infinite in

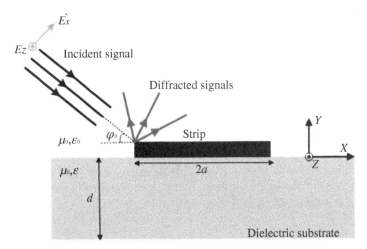

Figure 3.4 Geometry of diffraction by conductive strip on dielectric slab.

the z direction. The diffracted wave depends on the polarization of the plane wave. Two separate polarizations are considered: (i) Parallel polarization, $\vec{E} = E_0\vec{a}_z$, and (ii) orthogonal polarization, $\vec{E} = E_0\vec{a}_x$.

In the parallel polarization, the incident E-field is parallel to the main axis of the strip; the backscattered signal may have components in the x or z direction. The copolar and cross-polar components are defined by Equations (3.1) and (3.2), respectively [12]:

$$E_z^d \approx \sqrt{\frac{\pi}{2k_0\rho}} \exp\left[-jk_0\rho - j\frac{\pi}{4}\right]$$

$$\sum_{m=0}^{\infty} [A_m J_{2m}(k_0\cos\phi) + jB_m J_{2m+1}(k_0\cos\phi)]\tan\phi \qquad (3.1)$$

$$H_z^d \approx \sqrt{\frac{\pi}{2k_0\rho}} \exp\left[-jk_0\rho - j\frac{\pi}{4}\right]$$

$$\sum_{m=0}^{\infty} [C_m J_{2m}(k_0\cos\phi) + jD_m J_{2m+1}(k_0\cos\phi)]\tan\phi \qquad (3.2)$$

In the orthogonal polarization, in which the incident E-field is orthogonal to the polarization of the strip main axis, the copolar and cross-polar components are found from Equations (3.3) and (3.4), respectively:

$$H_z^d \approx \sqrt{\frac{\pi}{2k_0\rho}} \exp\left[-jk_0\rho + j\frac{\pi}{4}\right]$$

$$\sum_{m=0}^{\infty} [C_m J_{2m+1}(k_0 \cos \phi) + jD_m J_{2m+2}(k_0 \cos \phi)] \tan \phi \qquad (3.3)$$

$$E_z^d \approx \sqrt{\frac{\pi}{2k_0\rho}} \exp\left[-jk_0\rho + j\frac{\pi}{4}\right]$$

$$\sum_{m=0}^{\infty} [A_m J_{2m+1}(k_0 \cos \phi) + jB_m J_{2m+2}(k_0 \cos \phi)] \tan \phi \qquad (3.4)$$

In the above expressions, J_m is the Bessel function, (φ, ρ) is the coordinate of the observation point, and $k_0 = a \cdot \omega \sqrt{\mu_0 \cdot \varepsilon_0}$ is the normalized wave number. A_m, B_m, C_m, and D_m are the expansion coefficients and should be found separately by numerical approaches and through the following expressions [12]:

$$\sum_{m=0}^{\infty} [G(2m, 2n; k_0)][C_m + MA_m] = -M[J_{2n}(k_0 \cos \phi_0)] \qquad (3.5)$$

$$\sum_{m=0}^{\infty} [G(2m+1, 2n+1; k_0)][D_m + MB_m] = -Mj[J_{2n+1}(k_0 \cos \phi_0)] \qquad (3.6)$$

$$\sum_{m=0}^{\infty} [H(2m, 2n+1; k_0)][A_m - MC_m Z_0^2] = j \tan(\phi_0)[J_{2n+1}(k_0 \cos \phi_0)] \qquad (3.7)$$

$$\sum_{m=0}^{\infty} [H(2m+1, 2n+2; k_0)][B_m - MD_m Z_0^2] = -\tan(\phi_0)[J_{2n+2}(k_0 \cos \phi_0)]$$

$$(3.8)$$

where $n = 0, 1, 2, \ldots$, M is the admittance parameter and

$$G(\alpha, \beta; k_0) = \int_0^{\infty} \frac{J_\alpha(\xi) \cdot J_\beta(\xi)}{\sqrt{\xi^2 - k_0^2}} d\xi$$

$$H(\alpha, \beta; k_0) = \int_0^{\infty} \frac{J_\alpha(\xi) \cdot J_\beta(\xi)}{\xi} d\xi \qquad (3.9)$$

A comprehensive derivation of the above expressions based on the different values of the parameters has been provided in Ref. [12]. Among these results, the relation between the strip width and the cross-polar component is the concern. As shown in Ref. [12], the normalized wave number affects the level of the cross-polar component. Moreover, the cross-polar component is dependent on the angle of observation. For a certain value of the normalized wave number, that is relevant to the strip width, the maximum cross-polar component is achievable over a wide range of observation angles. The copolar and cross-polar components of the diffracted signal are figuratively shown in Figure 3.5. This figure shows that a well-designed strip is capable of creating significant cross-polar components while illuminated by a linear polarized incident signal.

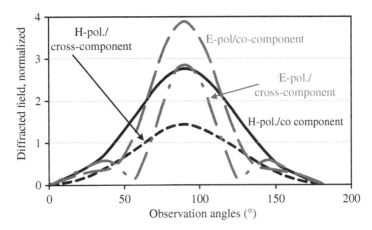

Figure 3.5 Copolar and cross-polar components of diffracted signals, linear scale.

This is an interesting result of diffraction. It simply means a strip with adequate dimensions may create noticeable cross-polar components in the backscattered signal. Hence, a simple strip line acts as an effective EM polarizer. This interesting aspect of the diffraction is used to suggest an effective data encoding algorithm in a chipless tag. One may recall that diffraction was the main issue in the barcode system as it limits the barcode's data encoding capacity and its reading range. In contrary, diffraction provides reliability and robustness in the proposed technique of imaging as will be shown.

3.3 STRIP-LINE POLARIZER

In the previous section, the cross-component of the backscattered signal from a narrow strip with infinite length due to the diffraction effect has been discussed. This simply means that the diffracted signal by the strip edge creates a cross-polar component. In addition to the diffraction effect, it is possible to maximize the cross-polar component of the backscattered signal by utilizing the resonance behavior of the strip-line structure. In a real scenario, the strip has a finite length; hence it is possible to design the length of the strip to resonate at a particular frequency. Maximizing the cross-component, one may suggest orienting the strip at 45° angle with respect to the incident plane wave. Therefore, the induced current on the strip surface, as a secondary source of EM wave, emits signal in both copolar and cross-polar directions as shown in Figure 3.6. It is important to note that in this scenario, the length of the strip plays a major role, while in diffraction phenomenon, the strip width is the main factor for cross-component creation.

The strip line shown in Figure 3.6 is simulated as an EM polarizer through a full-wave EM solver CST microwave studio. A linearly polarized plane wave illuminates the strip, while the incident E-field maintains 45° angle with the strip axis.

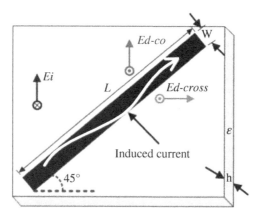

Figure 3.6 Strip orientation for creating cross-polar components, $L = 1.45$, $w = 0.2$, $\varepsilon = 2.55$, $h = 0.0127$, all in millimeters.

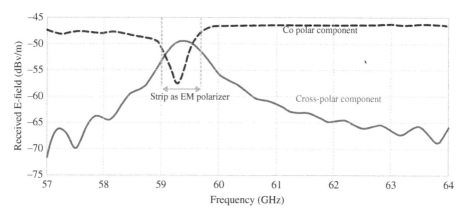

Figure 3.7 Simulated copolar /cross-polar components of backscattered signal, single strip line, $L = 1.45$, $w = 0.2$, $\varepsilon = 2.55$, $h = 0.0127$, all in millimeters.

Two E-field probes are oriented at a certain distance, 30 cm, for example, to measure the copolar and cross-polar backscattered signal. The results of the simulation are shown in Figure 3.7. The cross-polar component of the backscattered E-field shows peak and the copolar component shows null at the resonant frequency. This is due to the fact that the antenna (resonant) mode RCS cancels the structural mode RCS at the copolar direction [13]. In contrary, in the cross-polar direction, only the resonant mode is dominant; hence, the cross-polar component shows a peak value. This cross-polar component is immune to the copolar structural mode. As is clear from the figure, in a certain band depending on the strip length, the cross-polar component is significant and hence the strip line can be considered as an effective EM polarizer.

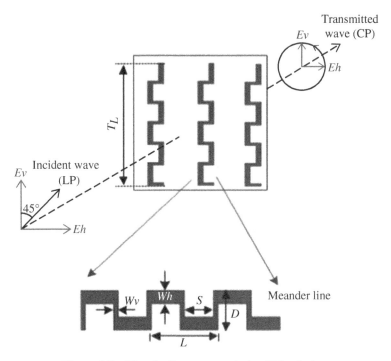

Figure 3.8   Meander line as transmission EM polarizer.

3.4   MEANDER-LINE POLARIZER

A meander-line polarizer consists of at least one layer of meander-line-shaped con-
ductors, usually printed on a circuit board [14]. A plane wave with 45° incident angle
with respect to the principal axis of the meander line illuminates the structure as
shown in Figure 3.8. The incident field can be decomposed into vertical and horizontal components. Each component experiences different phase shifts to provide
90° phase shift difference, while their amplitudes remain equal. As a result, the combined field in the transmitted signal is a circular polarized wave, and, therefore, the
component is called transmission EM polarizer [14] (see Figure 3.8). Normally, multiple layers of meander lines are required to create a complete circularly polarized
transmission signal [14–16]. A transmission meander-line polarizer also acts in the
reverse direction and can convert a circular polarized (CP) wave into a linear-type signal. This type of meander-line structure is widely used in satellite communications
[17,18].

   In the proposed application of the chipless RFID system, a reflection-type EM
polarizer is required as the reader normally measures the backscattered signal by the
tag. A possible solution for a reflection EM-polarizer structure is shown in Figure 3.9.
The structure utilizes multilayer meander lines with a ground reflector. The linearly
polarized incident wave (E_i) is converted into a circular polarization wave while

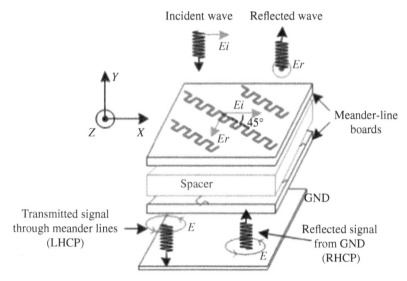

Figure 3.9   Multilayer meander-line structure for reflection EM polarizer.

passing the multilayer meander-line structure. For the x-polarized incident signal shown in Figure 3.9, the passing signal from the meander-line structure is a left-hand circular polarized (LHCP) wave, for example. This wave is completely reflected back by the ground plane while its rotation direction is reversed. Hence, the new wave with right-hand circular polarization (RHCP) is inserted in the multilayer meander-line structure again. This signal is converted into a linear polarized wave as the meander line changes the input circular polarized (CP) signal to a linearly polarized (LP) wave. However, the E-field of the final reflected wave (E_r) is orthogonal to the direction of the E-field of the original incident wave (E_i). Hence, the goal of the meander line as a reflection EM polarizer is achieved.

While the proposed approach of the multilayer meander line with a ground plane reflector is technically appropriate for the purpose of reflection-type EM polarizer, it would not be a suitable solution for chipless RFID tag structure. The multilayer tag structure contradicts with the expected subcent tag cost that is only addressed through a single-layer printable tag structure. Therefore, to propose a single-layer polarizer, the meander-line dimensions have been redesigned based on diffraction and resonance principles similar to those for the strip-line polarizer. This means that the parameters of the meander-line structure as shown in Figure 3.8 are designed to maximize the cross-polar component of the backscattered signal. Similar to the case of the strip line, again both diffraction and resonance behaviors of the meander line are used simultaneously for the maximum cross-polar component. The expectation is to achieve better performance from the meander line due to the increased design freedom. While the strip line has mainly two design parameters, strip width and length, the meander-line structure has more design parameters: W_h, W_v, D, S, and L.

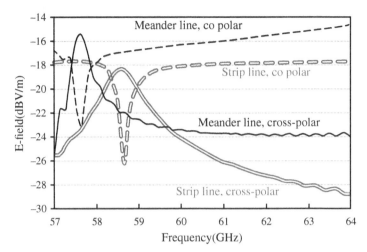

Figure 3.10 Simulated copolar/cross-polar component of single-layer strip line and meander-line structure.

A single-layer meander line is designed through parameter optimization in CST aiming to provide the highest cross-polar backscattered signal. The structure is assumed to have the same length as that for the strip line. The simulation result is shown in Figure 3.10, while the parameters of the design for the strip line and meander line are as follows:

For the strip line: $T_L = 1.6$, $\alpha = 0.2$

For the meander line: $T_L = 1.6$, $W_h = 0.2$, $W_v = 0.1$, $D = 0.37$, $L = 0.5$ (all values in millimeters), and $\varepsilon_r = 2.55$

It is clear from Figure 3.10 that the meander line's cross-polar component acquires sharper resonance for the same polarizer length ($T_L = 1.6$ mm). Moreover, the cross-polar E-field component of the meander line is 2–3 dB higher than that of the strip line. This equals to 5 dB more received power level. The higher cross-polar component sets freedom on the reading distance and better signal-to-noise ratio (SNR) of the system.

3.5 MULTIPLE POLARIZERS

The range and azimuth resolutions were introduced in Chapter 2. It was also shown that the fine range resolution of millimeter order is not possible to achieve due to the limited available frequency band. Therefore, the azimuth resolution is the only factor that can be used for providing the required spatial diversity for data encoding purpose in a chipless RFID system. This means that the detectable pixels on the tag surface are rectangle with few millimeters width and length of tens of centimeters order, as shown in Figure 2.10. Each rectangle independently encodes a certain amount of

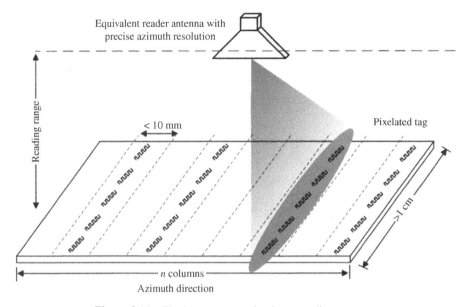

Figure 3.11   Final tag structure for data encoding purpose.

data. Considering the shape and dimensions of the suggested EM polarizer as encoding elements, it is expected that each rectangle may include multiple polarizers for increasing the encoding capacity if polarizers with different lengths and hence different resonance frequencies are used. Alternatively, one may use similar polarizers on each rectangle for higher RCS, which results in better SNR or a longer reading range. Hence, the final shape of the chipless tag for encoding purpose is shown in Figure 3.11. As multiple polarizers are suggested on each rectangle, it is important to investigate the effect of coupling among the polarizers. Process tolerance is also expected during tag fabrication; hence, it is very beneficial to explore the fabrication error on the final performance of the system. Doing these works, each scenario is first simulated and then compared with the measurement result.

3.5.1   Multiple Similar Strips, Coupling Effect

To find out the effect of coupling among polarizers on each rectangle, it is easier to start with multiple strips with the same dimensions. It is also beneficial to explore the effect of multiple similar polarizers on increasing the signal level for longer reading ranges or satisfying SNR.

   If an accurate and precise fabrication process is utilized for tag manufacturing, then it is reasonable to expect that all the strip lines have exactly the same dimensions. Obviously, if the strip lines are of the same length and width, then a similar response to what is shown in Figure 3.7, with a higher received E-field level, is expected. To validate this expectation, five strip lines with the same dimension are placed near to each other (1 mm distance). The simulated copolar and cross-polar backscattered

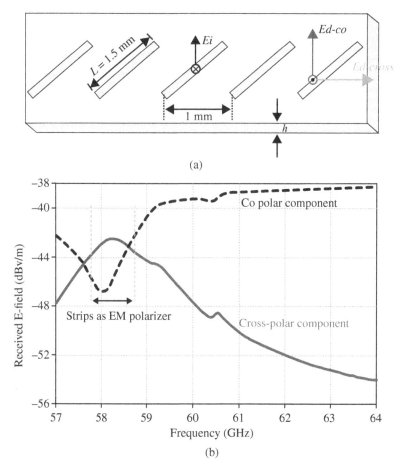

Figure 3.12 (a) Five strip lines with same length and (b) simulated copolar/cross-polar backscattered signals.

E-fields are shown in Figure 3.12. Comparing Figures 3.7 and 3.12, one notices the following phenomena:

(i) Resonance frequency shifts from 59.5 to 58.2 GHz, about 1.3 GHz.

(ii) Resonance bandwidth spread from 0.7 to 1 GHz.

(iii) Amplitude of the backscattered signal increases by approximately 6.5 dB from -49 to -42.5 dB/m. This is approximately five times ($10 \times \log 5 = 7$ dB) amplification as expected.

Both (i) and (ii) can be related to the coupling among strips as they were placed very close to each other (1 mm). This means that Figure 3.12 is not simply a multiplication of Figure 3.7. However, the initial expectation on the received power level has been almost confirmed.

3.5.2 Multiple Different Strips, Smeared Resonance

The effect of multiple strips with the same dimension has been considered in the previous section. It is useful to explore the system capability on providing higher data encoding capacity on each rectangle of the tag by combining multiple polarizers with different dimensions and hence various resonance frequencies. This means that the suggested approach of spatial diversity is combined with conventional frequency-signature technique for higher content capacity. Therefore, multiple polarizers, strip lines, with slightly different lengths are simulated.

Multiple strips with slightly different lengths are placed close (1 mm) to each other. Five strip lines, whose lengths vary from 1.6 to 1.7 mm, are considered. Figure 3.13 shows the simulation results. For ease of comparison, the result of a single strip is also reproduced in this figure. It is clear from Figure 3.13 that the response from multiple strips with different lengths does not show any resonance behavior. There are some fluctuations on the received signal level, but detection of the resonances is very difficult if not impossible. The mutual coupling as discussed is one of the factors. Moreover, one may suggest that each strip line is resonating at different frequency and hence the result is the combination of all resonances and hence no specific resonance is detectable. The importance of this result is revealed when a printed tag structure is considered.

3.6 POLARIZER FABRICATION

Proposed structures of strip line and meander lines as EM polarizers are fabricated for measurement purpose. First, polarizer-based tag is fabricated through an accurate photolithographic process on a dielectric structure. Then, the same tag structure is printed with a SATO inkjet printer that utilizes conductive ink on a paper substrate.

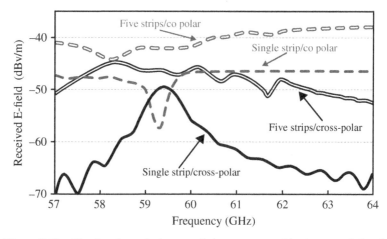

Figure 3.13   Simulated copolar/cross-polar components of backscattered signal.

These two tag samples through different fabrication processes are carefully investigated by an electronic microscope regarding their precise dimensions and shapes. It will be shown that the printed sample suffers severely from shape anomalies and changes in dimensions due to the printing resolution of the commercial printing technology.

3.6.1 Photolithographic Fabrication Process

The suggested polarizers in the previous section are designed and simulated based on the Taconic TLX-8 substrate with $\varepsilon_r = 2.55$ and 0.127 mm thickness. The conductive part is the annealed copper with 1 oz thickness. The design accuracy is 0.05 mm. It is possible to fabricate a single polarizer; however, as shown before, more than one polarizer are placed in each rectangle of the tag surface. The designed polarizers, strip line, and meander line are fabricated through a photolithographic process. The photographs of these two polarizers are shown in Figure 3.14.

3.6.2 Printed Polarizer

It was thoroughly discussed in the previous chapters that any suggestion for chipless RFID systems shall be based on the printed tag structure. Otherwise, obtained results based on perfect materials and costly fabrication processes do not guarantee the successful performance of any proposed approach at the commercial stage. This is the most significant phase of the work in each proposed technique for chipless RFID systems. To evaluate the performance of the proposed approach, the polarizers are printed by a SATO GL4e industrial thermal silver ink-based printer with a 600 dpi resolution [19]. SATO uses aluminum-based ink as the conductor and normal paper as the substrate. Figure 3.15 shows the SATO printer and the printed polarizers. The final printed structures are measured with an electronic microscope and the values are shown in the figure. The actual lengths of printed strip lines are normalized to show the difference among final printed structures and the designed lengths. For the meander-line case, the final dimensions are not shown due to the large number of design parameters for each object. Instead, the micrograph of a printed meander line is highlighted that clearly shows the anomalies occurring during the printing process.

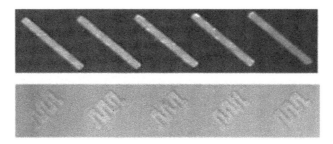

Figure 3.14 Fabricated EM polarizers through photolithographic process.

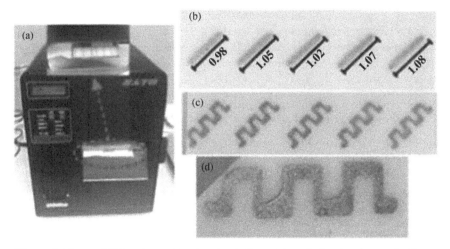

Figure 3.15   (a) SATO printer, (b) printed strip lines, (c), and (d) printed meander lines.

The effect of process errors on the polarizer's performance is shown in the following chapter when tags are measured in a real environment. However, it is possible to predict the effect of printing errors on the backscattered signal. Each printed strip line has a certain length and hence resonance at different frequency than what expected by simulation and design. This means that in real scenarios, it is not possible to have multiple printed strip lines with the same dimensions. Therefore, creation of multiple similar strip lines as discussed in Section 3.5.1 is not practically possible due to the limitation in printing accuracy of the commercial printing technologies. Instead, the situation discussed in Section 3.5.2, multiple different strip lines, is happening and so no specific resonances can be expected from a group of strip lines with slightly different dimensions. In conclusion, the combination of the frequency-signature technique with spatial diversity is actually not practical and the data encoding shall rely solely on the spatial diversity. This claim is confirmed by measurement in the following chapter.

3.7   CONCLUSIONS

Shaping of the tag surface for providing a unique image is discussed in this chapter. The similarities and differences between the earth imaging and tag imaging have been shown. It was found that due to the diffraction of EM wave from the edge of a strip line with proper dimensions, it is possible to expect the cross-polar component at the backscattered signal. This idea has been considered through KP approach. Then the effect of resonance response from an inclined strip line is added to the diffraction phenomenon to build the basis for the EM-polarizer base system. Meander line as a more effective EM polarizer has been introduced. The simulation results for a single polarizer and multiple polarizers were presented. The effect of multiple polarizers with slightly different dimensions on the cross-polar backscattered signal was then considered. It was found that a printed polarizer has inaccurate dimensions hence act similar to the multiple polarizers with different dimensions. This part is very important on proposing the correct approach for data encoding scenario in the proposed spatial-based system.

REFERENCES

1. JPL. (2013). Available: http://southport.jpl.nasa.gov/.

2. M.H. Williams, "Diffraction by a Finite Strip," *Oxford Journals, Mathematics & Physical Sciences,* vol. 35, 1980.

3. M.C. Heaton, P.J. Joseph, and R.L. Haupt, "Uniform Theory of Diffraction Analysis for Conductive Strips with Constant and Tapered Resistive Loads," in *Antennas and Propagation Society International Symposium,* 1991, p. 4.

4. U.Y. Ziya, "Physical Optics Theory for the Scattering of Waves by an Impedance Strip," *Optics Communications,* vol. 284, p. 1760, 2010.

5. R.G. Kouyoumjian and P.H. Pathak, "A Uniform Geometrical Theory of Diffraction for an Edge in a Perfectly Conducting Surface," *Proceeding of the IEEE,* vol. 62, 1974.

6. S.W. Lee, "Comparison of Uniform Asymptotic Theory and Ufimtsev's Theory of Electromagnetic Edge Diffraction," *IEEE Transactions on Antennas and Propagation,* vol. 25, 1977.

7. T. Griesser and C.A. Balanis, "Backscatter Analysis of Dihedral Corner Reflectors Using Physical Optics and the Physical Theory of Diffraction," *IEEE Transactions on Antennas and Propagation,* vol. 35, 1987.

8. R. Mittra, Y. Rahmat-Samii, and W.L. Ko, "Spectral Theory of Diffraction," *Applied Physics,* vol. 10, 1976.

9. V.C. Monk and F.W. Sedenquist, *"High-Frequency Radar Modeling,"* Advanced Sensors Directorate Research, Development, and Engineering Center, Alabama, 1995.

10. J.B. Keller, "One Hundred Years of Diffraction Theory," *IEEE Transactions on Antennas and Propagation,* vol. AP-33, p. 4, 1985.

11. A. Imran and Q.A. Naqvi, "Diffraction of Electromagnetic Plane Wave from a Slit in a PEMC Plane," *Progress In Electromagnetics Research vol.* 8, 2009.

12. M.A. Imran, "Diffraction of Electromagnetic Plane Waves from Strips and Slits Using the Method of Kobayashi Potential," PhD Thesis, Quaid-I-Azam University, Islamabad, 2010.

13. I. Balbin and N.C. Karmakar, "Phase-Encoded Chipless RFID Transponder for Large-Scale Low-Cost Applications," *IEEE Microwave and Wireless Components Letters,* vol. 19, pp. 509–511, 2009.

14. A.K. Bhattacharyya and T.J. Chwalek, "Analysis of Multilayered Meander Line Polarizer," *International Journal of Microwave and Millimeter-Wave Computer-Aided Engineering,* vol. 7, pp. 442–454, 1997.

15. L. Young, L.A. Robinson, and C.A. Hacking, "Meander-Line Polarizer," *IEEE Transactions on Antennas and Propagation,* 21, 1973.

16. R.S. Chu and K. Lee, "Analytical Model of a Multilayered Meander-Line Polarizer Plate with Normal and Oblique Plane-Wave Incidence," *IEEE Transactions on Antennas and Propagation,* vol. 35, p. 652, 1987.

17. M. Mazur and W. Zieniutycz, "Multi-layer Meanderline Polarizer for Ku-band," in *Microwave, Radar and Wireless Communications (MIKON-2000)*, p. 78.

18. F. Jian, "The Optimum Designing Method for Wide Bandwidth Meander-Line Circular Polarizer," in *Antennas, Propagation and EM Theory, ISAPE 2000*, pp. 10–13.

19. SATO Australia (September 2014). Available: http://www.satoaustralia.com/.

4

ATTRIBUTES OF EM POLARIZERS

4.1 INTRODUCTION

Electromagnetic (EM) polarizers were introduced in Chapter 3. The edge diffraction of the proposed polarizer and its antenna mode result in significant cross-polar components in the backscattered direction, which provides the basis for the data encoding in a chipless RFID system. In this chapter, the fabricated polarizers through the photolithographic process and printing technique are tested in laboratory environment. When the depolarization aspect of the suggested structure as polarizer is confirmed, then the cross-polarized working basis is reviewed in more detail. First, the equivalent system model is introduced to ease the network analysis of the proposed cross-polarized system. The effectiveness of the system for unwanted reflections is proved mathematically. Then, different measurement scenarios are considered to investigate the system robustness in various scenarios. The effect of reflections from other objects and multipath interferences are also considered in the measurement process. This finding suggests huge advantages of the EM-polarizer-based system for industrial environments. Moreover, the polarizer-based tag is shown to be very efficient for tagging of highly reflective items; water bottles or metallic cans, for example. The bending effect of the tag conformal to various types of objects, on its cross-polar performance, is also considered in this chapter. Finally, the effect of barriers on the system performance is investigated and shows an interesting aspect for the suggested technique, hidden tag-reading capability.

Advanced Chipless RFID: MIMO-Based Imaging at 60 GHz–ML Detection, First Edition.
Nemai Chandra Karmakar, Mohammad Zomorrodi, and Chamath Divarathne.
© 2016 John Wiley & Sons, Inc. Published 2016 by John Wiley & Sons, Inc.

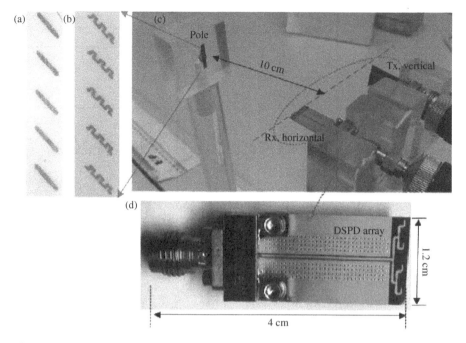

Figure 4.1 System measurement setup: (a) printed strip lines, (b) printed meander lines, (c) whole system, and (d) reader antenna.

4.2 SUGGESTED STRUCTURES AS EFFECTIVE EM POLARIZERS

Two structures, strip lines and meander lines, were suggested as EM polarizers and their performance was confirmed through simulation. It is beneficial now to validate their performance as polarizer in real scenario. For this purpose, a suitable measurement setup is required to maintain the cross-polar working basis.

The system setup for the measurement process is shown in Figure 4.1. On the reader side, two antennas are used. Antennas are double-side-printed dipoles (DSPDs) that have been designed to cover the whole suggested frequency band 57–64 GHz. They are very small and so are easily movable, due to the requirement for synthetic aperture radar (SAR). The main aspects of the antennas are their high axial ratios or equivalently low cross-polar level. This is vital for the cross-polar-based system. Moreover, antennas have a wide azimuth beamwidth that is required to interrogate the tag surface from different view angles. The design parameters and procedures of the antenna are discussed later. For measurement purposes, antennas are oriented orthogonally and they are connected to the performance network analyzer (PNA). Considering the low cross-polar level of the designed antennas and their orthogonal orientation, a very low signal leakage between transmit (Tx) and receive (Rx) antenna is expected. A plastic pole is installed in front of the antennas on which the tag is attached. The PNA's transmit power is −5 dBm, that is, its maximum power at the millimeter-band of 60 GHz. Due to the limitation of the PNA transmit power, the reading distance is set to 10 cm. It will be shown that the average gain of the antenna is 5.5–7 dBi, hence the maximum equivalent isotropically radiated power (EIRP) used in this measurement setup is only +0.5–2 dBm. As shown in Chapter 5, the maximum allowable EIRP at the suggested band is +40 dBm; hence, the reading distance in the proposed technique would increase up to 50 cm if adequate transmitter is available. The PNA measures the S-parameters versus frequency; hence, cross-RCS of the tag in dBm versus frequency is shown by the PNA.

4.2.1 Precise Polarizers

The fabricated tags through photolithographic process that have been introduced in Chapter 3 are attached to the pole for measurement purposes. The S_{21} in decibel versus frequency is shown in Figure 4.2. The dashed line shows the S_{21} when no tag is attached to the pole. The signal level is below −65 dBm. This presents the amount of signal leakage and mutual coupling between the two orthogonally polarized antennas as well as the effect of the plastic pole. The very low signal level is not unusual as the antennas are oriented orthogonally and each antenna has a low cross-polar level.

Attaching a tag comprises of five precisely fabricated strip lines through a photolithographic process significantly increases the received power level by at least 15 dB. The signal level would be up to 30 dB higher at the resonance frequency 60.2 GHz. This result clearly verifies the strip line as an effective EM polarizer. Comparing the measured results with the simulation outcomes in Chapter 3 confirms that the resonance bandwidth of the tag has been increased in a real scenario. The broader resonance bandwidth is important as it challenges the resonance-based idea for data encoding. Even though the tag has been fabricated through an accurate and

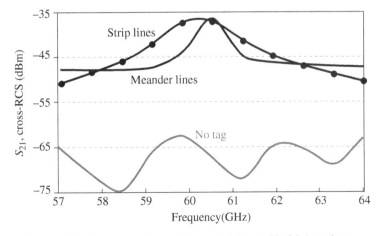

Figure 4.2 Measurement result for photolithographic fabricated tags.

expensive process by utilizing copper and high-quality dielectric substrate, the actual Q factor of the element is lower than that of expected in simulation.

Interrogating five precisely fabricated meander lines by the vertically oriented antenna creates a backscattered signal that includes a high amplitude cross-polar component. This signal is easily picked up by the horizontally oriented receive antenna. Similar to the striplines case, the received signal level is almost 18 dB higher than that of the "no tag" case. At the resonance frequency 60.5 GHz, the signal level is 30 dB stronger. The meander lines show a higher Q factor than that of the strip lines. However, their resonance bandwidth is again increased with respect to the simulation results shown in Chapter 3.

In summary, it can be concluded that the strip line and meander line act as an effective EM polarizer at the designed band 57–64 GHz. They both suffer Q factor reduction due to the nonideal behavior of elements in a real scenario, specifically at the 60 GHz millimeter-band.

4.2.2 Printed Polarizers

The printed tags using the SATO printer are considered for measurement purpose. The conductive aluminum-based ink is utilized over paper for the printing process. It has already been shown that the strip lines are not at the exact same lengths and meander lines suffer from structural anomalies occurring during the printing process. The measurement results are shown in Figure 4.3. For comparison purposes, the "no tag" case result is also shown. The received power level increases by at least 18 dB over the entire frequency range of 57–64 GHz if a printed stripline-based tag is attached to the reading pole. This clearly accredits the strip lines as EM polarizers even for a printed structure. However, in this scenario, no resonance behavior is seen in the response of the printed structure. Resonance(s) disappears due to the low conductivity of the

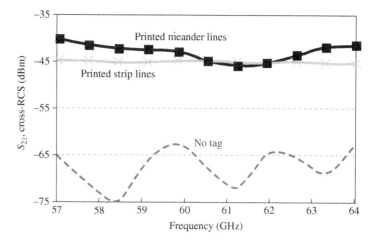

Figure 4.3 Measurement result for printed tags with SATO.

utilized ink and lossy paper substrate that significantly reduces the Q factor of the structure. Moreover, the printing inaccuracy results in strip lines with slightly different dimensions as shown in Figure 3.15. As expected by the simulation results, no resonance behavior shall be expected from their aggregate response.

Interrogating of the five printed meander lines also shows high cross-polar component. Similar to the striplines case, the received signal level is 20–25 dB higher than that of the "no tag" case. Moreover, as seen for the printed strip lines, no resonance is detectable at the whole frequency range of operation. Again, the low Q factors of the structure and shape anomalies due to printing inaccuracy are the reasons for almost uniform response on the 57–64 GHz band. The results of measurement for the printed EM polarizers are very important as they present the final shape of the tag in its commercial phase.

The results of measurement for the PCB and printed EM polarizers shall be considered very carefully as they present two critical aspects:

(i) *Proof of Polarizer Performance.* The results shown in Figures 4.2 and 4.3 confirm the performance of the strip and meander lines as effective polarizers. All measurements were performed in the laboratory environment and not in an anechoic chamber. Moreover, no calibration or reference tag has been utilized. These considerations endorse the suggested stripline and meander lines as suitable, low-cost, compact, and effective polarizers in the millimeter-band, which validates the data encoding method in the suggested image-based chipless RFID system.

(ii) *Rejection of Resonance-Based Encoding Approaches.* Extensive system tunings have been considered to calibrate the SATO printer for the most accurate printing result. All the printed tag structures show almost the same trend in varying the lengths and dimensions. It appears that the low-cost printing technology is not mature enough for the demanded accuracy in RFID applications. This suggests that any resonance-based approach for data encoding faces huge limitations at the commercial phase of the system implementation. High-resolution printing methods in the order of 10–20 μm are too expensive and slow processes, laser printing, for example.

Irrespective of lost resonant at the backscattered signal due to the printing inaccuracy and low quality of materials in the tag structure, the measured high cross-polar signal can be utilized for a cross-polar RCS-based data encoding algorithm. This new basis is used in the proposed theory of the EM-image chipless RFID system. Therefore, in the proposed image-based technique, the system does not rely on resonance detection; hence, it is robust toward the printing inaccuracy, and, therefore, the low-cost printers are still viable methods.

4.3 CROSS-POLAR WORKING BASIS

After confirming that the printed polarizers are able to provide cross-polar working basis for the proposed technique, it is very beneficial to explore the advantages of the cross-polar rather copolar. First, the equivalent system model is introduced to ease the network analysis of the proposed EM-polarizer-based system. The effectiveness of the system for unwanted reflections is proved mathematically. Then, different measurement scenarios are considered to investigate the system robustness in various scenarios.

4.3.1 Analytical Model

To explore the potential of the cross-polar-based system, the equivalent system network is introduced. This enables the mathematical analysis of the tag response and realization of the system robustness toward reflection and multipath interferences. Moreover, in the theoretical evaluation of the proposed technique, the drawbacks and limitations of the copolar-based system as the working basis of many current techniques in the chipless RFID systems are explored. It should be noted that the frequency, time, and most of hybrid domain-based chipless RFID tags work on the copolar backscattered signal, hence susceptible to noise and interferences. To alleviate the situation, rigorous calibration is required to extract data.

As previously discussed, the tag is interrogated with a linearly polarized signal, for example, vertical polarization. The backscattered response is picked up by the receiver in the orthogonal direction. The RCS of the tag relates to many parameters including the frequency and polarization of the impinged signal, the aspect angle of the tag, and its physical size and structure [1]. For a certain frequency and aspect angle, the RCS behavior depends on the interrogation signal's polarization by Wiesbeck and Kahny [2]:

$$\sigma = \begin{bmatrix} \sigma_{vv} & \sigma_{vh} \\ \sigma_{hv} & \sigma_{hh} \end{bmatrix} \tag{4.1}$$

In Equation (4.1), σ represents the tag's scattering behavior. Each matrix element, σ_{ij}, shows the tag response in a j polarization scheme while interrogated by an i polarized signal. For simplicity, the vertical (v) and horizontal (h) polarizations are assumed in this work, while Equation (4.1) is valid for any orthogonal polarizations.

Evaluation of the proposed technique requires a circuit model. The suggested model shall include all the elements of the technique. One possible model for the chipless RFID system was introduced by Vena et al. [3]. This model includes the effect of Tx and Rx antennas, the response of the chipless tag, and the equivalent response of all interfering objects surrounding the tag as shown in Figure 4.1. Each block in the model shows a specific section of the communication link. As two vertical and horizontal polarizations were assumed, then each box has a 2×2 S-parameters matrix that characterizes its behavior in co- and cross-polar directions.

In Figure 4.4, the transceiver module on the reader side is denoted by block M. The source is assumed to send a pure V-polarized signal and the receiver is expected to collect the backscattered signal only in H-polarization direction. However, due to the nonideal behavior of the elements involved in the link, there is the possibility of errors in the tag detection process. The model is intended to find those errors and mathematically evaluate their weights in the final system performance. The Tx antenna and its connected cable and connectors are symbolized by block T or the transmit path. Similarly, the receiver antenna and cables are shown through box R

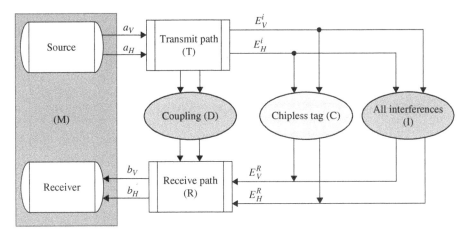

Figure 4.4 Schematic block diagram of the chipless RFID system.

or receive path. The coupling among Tx and Rx antennas always exists. To show the effect of coupling, block D is considered in the model. Existence of the tag in front of the antennas causes reflection of energy toward the reader's receive antenna. This is the intended reflection that carries the tag information. This link is shown by block C in Figure 4.4. Apart from direct coupling and reflection by the tag, there is the possibility of receiving the signal through reflections by other objects around the tag. This includes everything located in the reader's reading zone. The item to which the tag is attached such as the walls, floor, and ceiling are examples. Normally, these reflections are undesirable phenomena and known as interference. Their effects significantly contribute to system performance and, therefore, shall be considered in the schematic block diagram. Box I represents the effect of all those interfering factors, so it covers all unwanted reflections, multipath interferences, and clutter.

The S-parameter matrix of the reader, M, can be related to the S-matrix of other blocks by

$$M = TDR + TCR + TIR \qquad (4.2)$$

In this equation, TDR shows the amount of coupling or direct leakage between transmit and receive antennas. This part is independent of the effect of the tag or other objects. Therefore, leakage, L, can be directly measured when there is nothing in front of the antennas.

$$L = TDR \qquad (4.3)$$

Therefore, one can combine Equations (4.2) and (4.3) to have

$$\begin{bmatrix} M_{vv} & M_{vh} \\ M_{hv} & M_{hh} \end{bmatrix} = \begin{bmatrix} L_{vv} & L_{vh} \\ L_{hv} & L_{hh} \end{bmatrix} + \begin{bmatrix} T_{vv} & T_{vh} \\ T_{hv} & T_{hh} \end{bmatrix} \cdot \begin{bmatrix} C_{vv} & C_{vh} \\ C_{hv} & C_{hh} \end{bmatrix} \cdot \begin{bmatrix} R_{vv} & R_{vh} \\ R_{hv} & R_{hh} \end{bmatrix}$$
$$+ \begin{bmatrix} T_{vv} & T_{vh} \\ T_{hv} & T_{hh} \end{bmatrix} \cdot \begin{bmatrix} I_{vv} & I_{vh} \\ I_{hv} & I_{hh} \end{bmatrix} \cdot \begin{bmatrix} R_{vv} & R_{vh} \\ R_{hv} & R_{hh} \end{bmatrix} \qquad (4.4)$$

As already assumed, the transmit signal is V-polarized and the receive signal is H-polarized. Therefore, M_{vh} is the interested S-parameter (forward transmission coefficient) for the proposed cross-polar system. M_{vv} is also important to compare the performance of the proposed system with other copolar-based systems. These two S-parameters are driven by

$$M_{vv} = L_{vv} + (T_{vv}C_{vv} + T_{vh}C_{hv})R_{vv} + (T_{vv}C_{vh} + T_{vh}C_{hh})R_{hv}$$
$$+ (T_{vv}I_{vv} + T_{vh}I_{hv})R_{vv} + (T_{vv}I_{vh} + T_{hv}I_{hh})R_{hv} \qquad (4.5)$$

$$M_{vh} = L_{vh} + (T_{vv}C_{vv} + T_{vh}C_{hv})R_{vh} + (T_{vv}C_{vh} + T_{vh}C_{hh})R_{hh}$$
$$+ (T_{vv}I_{vv} + T_{vh}I_{hv})R_{vh} + (T_{vv}I_{vh} + T_{hv}I_{hh})R_{hh} \qquad (4.6)$$

Equations (4.5) and (4.6) are related to the physical characteristics of the elements in the reading process. If they are carefully investigated, the complexity of Equations (4.5) and (4.6) can be reduced. For example, the terms I_{vh} and I_{hv} represent the depolarization aspect of the unwanted reflection toward the reader. The usual expectation is that reflection through surrounding objects would be at the same polarization. Specifically at millimeter-wave range, the objects are too big to create a depolarized signal due to the diffraction effect. Multipath interferences and clutter also show no depolarization behavior, hence I_{vh} and I_{hv} can be correctly assumed to be zero. T_{ij} and R_{ij}, on which $i \neq j$, are related to the antenna polarization or the antenna's CPL. As shown in the following chapter, the designed array antenna for the proposed image-based chipless RFID system has a high CPL of more than 20 dB. Moreover, the PNA Agilent E8361A has a high-quality cable and connectors with very high isolation in the order of 80–100 dB [4]. This means that the value of T_{ij} and R_{ij} are less than 0.01 and again can be correctly neglected.

By applying the aforementioned simplifications, Equations (4.5) and (4.6) become

$$M_{vv} = L_{vv} + T_{vv}R_{vv}(C_{vv} + I_{vv}) \qquad (4.7)$$

$$M_{vh} = L_{vh} + T_{vv}C_{vh}R_{hh} \qquad (4.8)$$

In these formulas, L_{ij} refers to the leakage between Tx and Rx modules of the reader whether in co- or cross-polar directions. The following statements are related to the leakage:

- *Leakage Measurement.* In an empty environment, leakage (L_{ij}) can be directly measured. The value of L_{ij} depends on the internal characteristics of the designed antenna and is also related to the physical distance between the Tx and Rx antennas. Once L_{ij} is measured, it does not change with varying the reading scene.
- *Significance of Leakage.* The leakage for the copolar case would be much higher than that of the cross-polar scenario $L_{vv} > L_{vh}$. If the copolar leakage

(L_{vv}) is not controlled well, it may significantly affect the final system performance.

• *Leakage Control.* The leakage control in the copolar scenario, L_{vv}, is much more complex than that for the cross-polar, L_{vh}. This is because in the copolar case, both transmit and receive antennas are on the same polarization scheme and, hence, they are more likely to interfere with each other.

Therefore, it can be concluded that the leakage in the cross-polar case is much less than that of the copolar and, hence, it has less effect on the reading process. Moreover, the leakage control is also much easier in the cross-polar scenario. This is one of the most significant advantages of the cross-polar scenario over the copolar working basis.

In Equation (4.8), which relates to the cross-polar scenario, the term $T_{vv}C_{vh}R_{hh}$ refers to the mixed effects of the Tx, Rx, and the tag. To accurately determine the coupling coefficient $T_{vv}.R_{hh}$ due to the filtering effect of the transmitter and receiver antennas, it is required to utilize a reference tag with known scattering behavior. Therefore, the tag's scattering term (C_{vh}) can be easily found as

$$T_{vv}R_{hh} = \frac{M_{vh}^{ref} - L_{vh}}{C_{vh}^{ref}} \tag{4.9}$$

$$C_{vh} = \frac{M_{vh} - L_{vh}}{M_{vh}^{ref} - L_{vh}}C_{vh}^{ref} \tag{4.10}$$

where L_{vh} is a fixed value based on the Tx and Rx antennas' distance and M_{vh}^{ref} is the measured scattering parameter of the reference tag.

One may suggest performing the same approach for the copolar scenario. However, as it is clear from Equation (4.7), an extra term I_{vv} also exists for the copolar case. The I_{vv} counts the effect of reflections from surrounding objects around the tag. As the tag size is very small compared to other surrounding objects, the I_{vv} is much larger than the C_{vv} in Equation (4.7). This means that it is not possible to extract the tag response without the information about its surrounding objects. Indeed, it is required to first calibrate the system based on a specific orientation of objects in the reading zone and then detect the response of the tag by subtracting two measurement results. Although with this calibration process, the tag reading would be possible in the copolar scenario, the approach is very sensitive to any changes in the surrounding objects' orientations or tag movement. This means that if the tag moves in the reading zone or if one object around the tag changes its place or aspect angle, then it is required to calibrate the system again. On the contrary, the cross-polar scenario is not dependent on the surrounding objects and no information is required. This is a very significant aspect of the cross-polar working basis, as the system is working well in an unknown reading zone only with a unique calibration setup. This aspect of the proposed cross-polar system is shown in a practical scenario in the following sections.

4.3.2 Multipath Interference and Clutter

Interferences due to the reflections from objects around the tag are a major problem in many conventional chipless or even chip-based RFID systems. The analytical model of the system's performance regarding the multipath interference and clutter was shown for the co- and cross-polar systems. Here, the system is tested in a real environment to validate the mathematical expectation. The measurement setup is similar to that shown in Figure 4.1. The reading distance, transmit output power, and all other parameters are the same for comparison purpose. Exploring the effect of multipath and reflection interferences on the reading process, the measurement is accomplished in the lab environment and not in an anechoic chamber. Moreover, a harsh and severe multipath situation is implemented by using some randomly selected highly reflective objects. Three horn antennas are directed toward the reader antennas. A large printed circuit board with $35 \times 45 \, \text{cm}^2$ is also placed in front of the reader just after the reading pole as shown in Figure 4.5.

To simulate the high-clutter situation, a big metal plate is attached to the reading pole. The size of the plate is very large, $20 \times 25 \, \text{cm}^2$, compared with the reader antennas and the reading distance (Figure 4.5). This highly reflective barrier causes a challenging situation for many traditional chipless RFID systems. The reflection from this plate (I_w) is much higher than that of the tag (C_{vv}) as discussed in the previous section. Therefore, the detection of the tag would be impossible unless the effect of high reflection is deleted through a calibration process. The calibration, however, is sensitive to any changes in the orientation of the objects around the tag and the tag movement.

The result of measurement for these two scenarios is shown in Figure 4.6. For comparison purposes, two results from normal measurement process, a "no tag" case and printed polarizers, are also repeated in Figure 4.6. As the graphs show, the effect of multipath interference and clutter on the received signal are almost negligible and very similar to the case when no tag is attached to the pole. This clearly matches with the analytical conclusion regarding the robustness of the cross-polar system to multipath and clutter interferences. Reflections from objects around the tag are mostly copolar, while the receiver picks up cross-polar components only. This significant advantage of the proposed cross-polar-based technique is very important in practice with respect to other copolarized-based systems [5–7]. In most cases, chipless RFID systems are vulnerable to multipath interferences and they require supplementary approaches, for example, multistep calibration or reference tags [6–9].

4.4 EFFECT OF HIGHLY REFLECTIVE ITEMS

In the previous section, the robustness of the proposed technique toward severe multipath interferences and high-clutter situations has been verified. This section aims at evaluating the performance of the printed polarizers when they are attached to highly reflective items. This covers the ultimate practical scenarios of the chipless

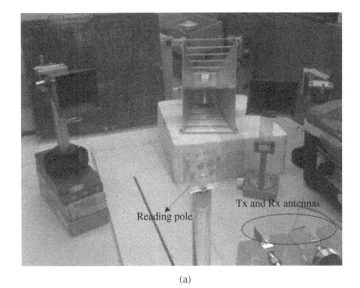

(a)

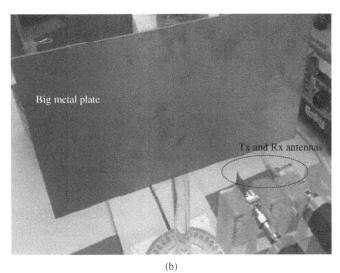

(b)

Figure 4.5 (a) Severe multipath interference scenario and (b) high clutter for the reading process.

RFID systems. For this purpose, two highly reflective objects are considered. First, a plastic bottle containing water is measured. As the reflection from water is unpredictable, so it may cause degradation of the reading process. Second, an aluminum can containing liquid is considered. The high reflection from the conductive body of the container as well as the bending effect affects the measured results.

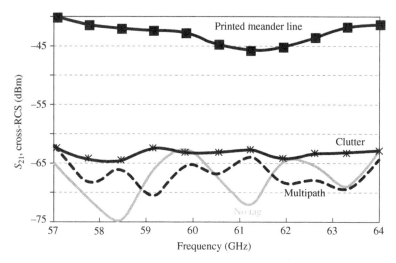

Figure 4.6 Measurement result for multipath and high-clutter situations.

4.4.1 Liquid Container

A normal plastic bottle of water is considered for the measurement. The tag is a single-layer structure consisting of printed polarizers on paper. The object and attached printed tag are shown in Figure 4.7. The tag consists of normal printed meander lines as highlighted in the picture. The measurement result is also shown in Figure 4.7. For comparison purposes, the measured results of the "no tag" case are also shown in Figure 4.7. As the graphs show, when no tag is placed in front of the antennas, the received signal level is below −65 dBm. Having a water bottle with no attached tag that is relatively very large has a minor effect on the cross-polar component. A maximum of 5 dB increase is experienced in this case. However, if the same water bottle includes a tiny printed tag as shown in Figure 4.7, then the received signal is significantly elevated. The received signal level from the attached tag is 12 dB higher compared to when the water bottle has no attached tag. This margin provides a reliable reading scenario of the liquid containers. Therefore, it can be concluded that the system has satisfactory performance when attaching to the liquid container. It is important to mention that no calibration process has been used in obtaining the graphs of Figure 4.7. Obviously, with a calibration process, much higher signal level differences are expected.

4.4.2 Metallic Objects

Tagging of metallic objects is always a challenging issue for RFID technologies due to the high reflection effect. An aluminum can is considered to evaluate system performance for this scenario. The tag is an etched-out meander line of conductive surface on paper. This means that the base of the tag is conductive and the meander line shape is paper. Therefore, this tag is a negative image of the former tag used in the previous step for the water bottle. The measurement setup and results are shown in Figure 4.8

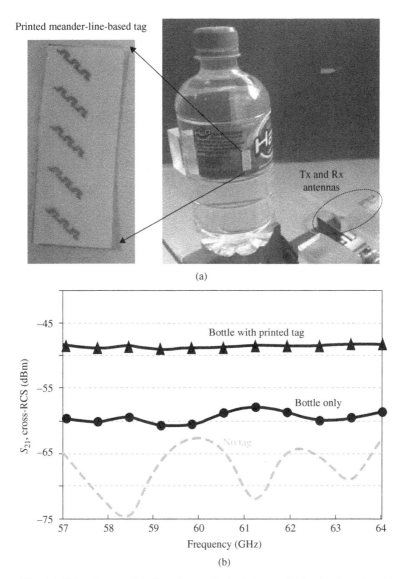

Figure 4.7 (a) Printed meander line tags attached to a plastic bottle of water and (b) measurement result.

with the highlighted tag structure. The received signal level for the "no tag" case is also depicted in this figure.

An aluminum can containing liquid without any attached tag, if placed in front of the reading pole, increases 5–7 dB of the receive signal in the cross-polar direction. If the same object is accompanied by a tiny printed tag, then it creates at least 13 dB higher cross-RCS on the reader side compared with the "can only" case. It is again

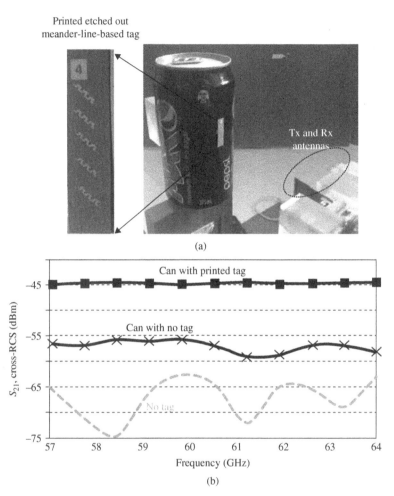

Figure 4.8 An etched-out printed meander line tag attached to an aluminum can and (b) measurement result.

important to indicate that all the graphs in Figure 4.8 are the pure measured signal level without any calibration or cancelling out of interferences process. This measurement result clearly shows the potential of the proposed technique for tagging of metallic objects.

4.5 SECURE IDENTIFICATION

One of the limitations of the barcodes is the requirement for direct code reading, for example, barcodes shall be exposed to the items and it is not possible to hide optical labels inside the objects. On the other hand, however, chipped RFID systems normally provide the possibility of hiding the tags inside the items, hence suggesting secure identification. It is therefore an utmost benefit to verify the possibility of secure identification in the suggested spatial-based technique. This means that the polarizer-based tag is readable in no "line-of-sight" (LoS) condition. Before exploring this aspect of the system, it is important to clarify the LoS terminology in EM theory. The LoS normally refers to propagation of EM wave including optics in a straight line. Based on the frequency of operation, a radiocommunication link may be referred to as LoS or NLoS. Low-frequency EM wave does not need an LoS condition, HF broadcasting, for example. For higher bands, the transmitter shall see the receiver to be able to communicate effectively or an LoS condition is required, point-to-point communication in gigahertz bands, for example. Based on this assumption, the proposed technique of imaging system is obviously an LoS system. Therefore, a straight path between the reader and tag is required for data encoding purpose.

However, the LoS systems may be categorized further. The communication link in some LoS systems may require a clear/visible LoS between Tx and Rx without any obstacles. Very-high-frequency EM wave and specifically optical communication systems fall into this category. These systems may be referred to as optical LoS systems. However, there are LoS systems that operate effectively even if some obstructions, in the form of EM-wave transparent materials, are in the reading zone. To categorize these two systems, the word optical LoS is used for the first systems. The second type of systems may be referred to EM-LoS system. The optical-based identification systems, barcodes and QR codes, are definitely optical LoS systems. Any barrier, even dirt, may result in reading failure. This restricts the system flexibility and is seen as a main issue for many applications.

Considering the mm-wave band of operation, it is important to evaluate the system performance when the tag is read in None-optical LoS condition, tag in an envelope or inside a box as example. This shows the system performance in None optical LoS conditions. For this purpose, multiple barriers are placed between the tag and reader. In this process, the tag is covered by the medium. This is similar to the cases when the tag is attached to an item and covered by a medium as shown in Figure 4.9. The medium could be an envelope paper or a box made up from cardboard or plywood. Table 4.1 shows different types of media, their thickness, and material compositions. This selection may cover a wide range of items.

The measured result and the images of different types of media are shown in Figure 4.10. For all cases, except for a tick wood (2 cm thickness), the system is adequately capable of detecting the presence or absence of the tag from the different measured signal levels in the cross-polar RCS of the tag. The millimeter-wave signals are very lossy in those media (plywood, cardboard, etc.) and mainly require a clear line-of-sight reading condition. However, as the system is cross-polar basis, the receiver is capable of detecting even a weak signal due to the natural cancelation of background noise. This provides a substantial advantage for the reading process. This simply means that in practical scenarios, the tagged objects can be packaged without affecting the data decoding process. It is important to emphasize that the results

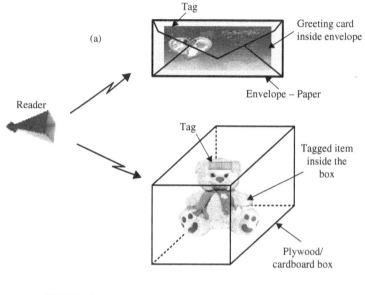

Figure 4.9   (a) NLoS reading scenario and (b) measurement process.

TABLE 4.1   Multiple Barrier

Barrier Name	Thickness (mm)	Material
Bubble envelope	4	Paper and plastic
Foam	18	Polypropylene/polyethylene (EPP/EPE)
Cardboard	2	Torn cardboard
Plywood	3	Wood veneer
Wood	20	Wood

in Figure 4.10 are without any calibration process. Obviously, if the environmental effects are cancelled out with a predefined calibration process, then much cleaner signal level differences between the two cases are observed. This process may ease reading of the tag through 2-cm-thick wood as well.

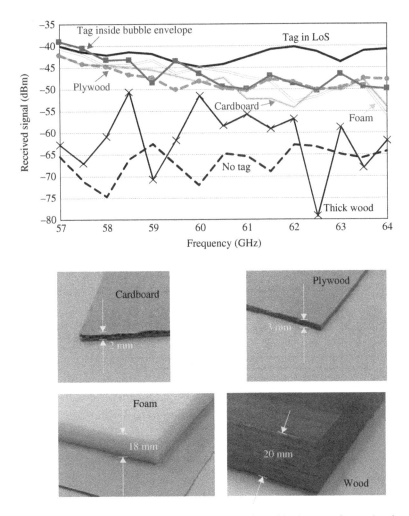

Figure 4.10 Measurement results for NLoS scenarios with pictures of some barriers.

4.6 BENDING EFFECT ON TAG PERFORMANCE

In real identification and tracking applications, the attached tags, whether barcodes or RFID types, are vulnerable to bending effect. This means that when a tag is attached to an item, the tag may not necessarily be in a flat configuration depending on the shape of the item. This issue affects the performance of the tag in the decoding process. As the tag size increases, degradation of the system performance due to the bending effect is becoming a serious issue. There are very limited published works on the bending factors for RFID tags or even barcodes [10]. The proposed tag in the spatial-based technique is physically a miniaturized item because of its millimeter-wave frequency band. Consequently, the expectation is that bending would not be a serious issue for a wide range of products to which the proposed tag can be attached.

Therefore, the system may provide a good robustness toward bending due to its miniaturized nature. However, it is important to explore how tag bending may affect the reading process in a real scenario if the items are very small compared to the tag size.

Moreover, it is important to clarify the meaning of bending. The final tag structure is a narrow strip with a fixed width around 1–1.5 cm and the tag length varying from 5 to 10 cm depending on the demanded data capacity (Figure 4.11). Consequently, bending is expected to occur along the tag length and not on its width as highlighted in Figure 4.11.

To verify the performance of the proposed tag with respect to bending, small paper rolls of varying radii are prepared to which a tag can be attached. These paper tubes are denoted as S-i; $i = 1, 2, 3, 4$. Figure 4.12 shows the tubes along with their radius. For comparison purposes, a pencil is also included in Figure 4.12. To better categorize

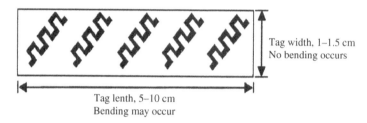

Tag width, 1–1.5 cm
No bending occurs

Tag lenth, 5–10 cm
Bending may occur

Figure 4.11   Expected bending on tag length.

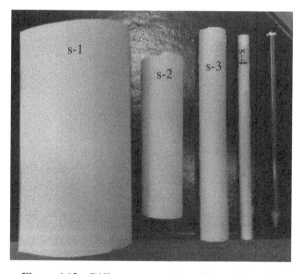

Figure 4.12   Different paper tubes with varying radii.

TABLE 4.2 Dimensions of Paper Tubes and Some Well-Known Items

Attached Items	Curved Radius, R (cm)	Arc Angle (°)
Paper tubes S-1	2.5	34
Paper tubes S-2	1.25	68
Paper tubes S-3	0.85	101
Paper tubes S-4	0.35	245
Aluminum can	3.25	26
Water bottle	2.75	31
Pencil	0.4	214

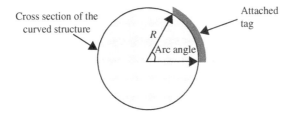

Figure 4.13 Relation between arc angle of attached tag and radius of objects.

the paper tubes, their radii along with the radii of some commonly used objects, such as a water bottle, an aluminum can, and a pencil, are summarized in Table 4.2. The second column of the table shows the curved radius of the objects to which the tag is attached. As can be seen, the radius of the biggest paper roll is smaller than that for a water bottle. The arc angle of the tag relates to the radius of the object to which the tag is attached. This arc angle also appears in Table 4.2. Figure 4.13 figuratively shows how the object radius is related to the arc angle of the attached tag. As the length of the attached tag is fixed, the bigger objects create less arc angle of the attached tag.

Figure 4.14 shows the measurement setup for the bending effect purpose. Again two separate antennas are utilized as Tx and Rx that are located at 10 cm distance to the reading pole and antennas are connected to the PNA. The measured cross-polar components of the backscattered signal from various curvatures are depicted in Figure 4.15. For objects bigger than S-1($R = 2.5$ cm), bending has negligible effect on the received power level. Examples include a water bottle and aluminum can. This clearly confirms system robustness for a wide variety of objects. This simply means that any object bigger than a water bottle experiences no bending effect on the proposed technique. For objects whose radii are $R < 2.5$ cm, the received power level gradually decreases. However, the receiver is still able to separate the signal of a bent tag from a "*no tag*" case. Only for items smaller than S-4 (0.35 cm radius), such as a pencil, bending significantly degrades the system performance. For those items, the received signal level is only 5 dBm higher than that of the "no tag" case, which may create uncertainty in the reading process. From a result, it can be concluded that

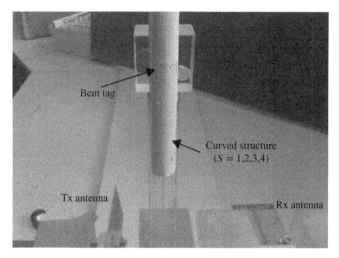

Figure 4.14 System structure for bending effect measurement.

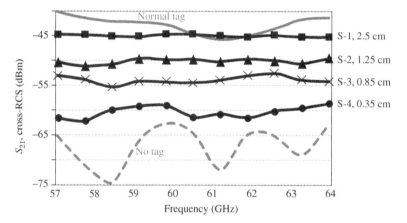

Figure 4.15 Measurement result based on different tag bending scenarios.

items with a curvature radius of more than 1.25 cm have an acceptable performance when the tag is attached in the proposed technique.

4.7 CONCLUSION

In this chapter, the suggested structure for the EM polarizers was processed based on different approachs to PCB and printing technique. The final tags in each approach were measured in real scenarios. While the precisely fabricated tags show almost similar response as predicted by simulation, the printed tag structure totally lost the resonance behavior. Irrespective of nonresonant response, the printed tags were still able to create significant cross-polar response in the backscattered signal, hence act appropriately as polarizer.

Then an analytical model of the reading scenario was developed based on radar theory. In the proposed equivalent network of the system, the effects of Tx and Rx antennas as well as the connecting cables and connectors were considered. Then the tag was considered in the communication channel between Tx and Rx antennas. Multipath and reflection from objects around the tag were also addressed in the analytical model. The system was mathematically analyzed and it was shown that the system is very robust to multipath interferences.

Moreover, it was shown that copolar-based systems are very vulnerable to changes in the reading zone and they require frequent system calibration. In contrast, the proposed cross-polar system only involves one single calibration process and the tag movement or other changes in the reading environment do not affect system performance. This issue was nominated as a significant advantage of the proposed technique compared to other current approaches in RFID applications, which are normally copolar based.

Different measurement scenarios were considered for confirmation of the analytical model of the system. First, a severe multipath situation and highly cluttered condition were considered and the tag response confirmed the system robustness as predicted by the analytical model. Then a printed meander-line-based tag was attached to a bottle of water. Moreover, an etched-out printed meander-line-based tag was considered as the negative image of the normal tag for attaching to an aluminum can. In both cases, the tag-reading process was successfully accomplished. While the effects of the water bottle and aluminum can without any attached tag were almost negligible in the cross-polar direction, the presence or absence of an attached printed tag could be easily detected by the reader system in a real scenario. This step confirmed a noticeable potential of the proposed theory with respect to other techniques in the RFID applications. Most proposed approaches in a chipless RFID system normally face difficulty in reading the tag in the presence of highly reflective items.

Next, the system was tested for NLOS conditions. Different types of barriers were placed between the tag and reader and then the system performance was measured. It was shown that the reader easily detects the presence or absence of printed polarizers for a wide range of barriers including, plywood, cardboard, and paper. This suggests huge potential for the proposed system on reading the tagged objects with an envelope or inside the box without the requirement for opening them.

In the last part of this chapter, the effect of tag bending is considered. The proposed tag structure showed a very good performance for a bent tag. This was initiated by the millimeter-wave band of operation, which resulted in miniaturized tag size with very low bending vulnerability. Only for objects that are very small and narrow, a pencil, for example, the bending may affect the reading process in the proposed chipless system.

The very successful performance of the proposed meander line as an effective EM-polarizer in harsh environments suggests a very robust system. Considering that all the measurements were accomplished in a real laboratory environment using a poorly printed tag, it appears that the proposed system for chipless RFID systems is fully practical for industrial applications.

REFERENCES

1. M. Skolnic, *Radar Handbook*, Third ed. McGraw-Hill, 2008.
2. W. Wiesbeck and D. Kahny, "Single Reference, Three Target Calibration and Error Correction for Monostatic Plarimetric Free Space Measurements," *Proceedings of the IEEE*, vol. 79, pp. 1551-1558, 1991.
3. A. Vena, E. Perret, and S. Tedjni, "A Depolarizing Chipless RFID Tag for Robust Detection and Its FCC Compliant UWB Reading System," *IEEE Transactions on Microwave Theory and Technique*, vol. 61, pp. 2982-2995, 2013.
4. Agilent Technologies. (Oct 2014). *Vector Network Analyser*. Available: http://www.agilent.com.
5. I. Jalaly and I.D. Robertson, "RF barcodes using multiple frequency bands," in *International Microwave Symposium Digest, IEEE MTT-S* 2005.
6. C.M. Nijas, R. Dinesh, U. Deepak, A. Rasheed, S. Mridula, K. Vasudevan, *et al.*, "Chipless RFID Tag Using Multiple Microstrip Open Stub Resonators," *IEEE Transactions on Antenna and Propagation*, vol. 60, pp. 4429-4432, 2012.
7. Y.F. Weng, S.W. Cheung, T.I. Yuk, and L. Liu, "Design of Chipless UWB RFID System Using A CPW Multi-Resonator," *IEEE Antenna and Propagation Magazine*, vol. 55, pp. 13–31, 2013.
8. F. Costa, S. Genovesi, and A. Monorchio, "A Chipless RFID Based on Multiresonant High-Impedance Surfaces," *IEEE Transactions on Microwave Theory and Techniques*, vol. 61, 2013.
9. N. C. Karmakar, *Handbook of Smart Antennas for RFID Systems*. John Wiley, Singapore, 2010.
10. J. Siden, P. Jonsson, T. Olsson, and G. Wang, "Performance degradation of RFID system due to the distortion in RFID tag antenna," in *Microwave and Telecommunication Technology*, Sevastopol, Ukraine, 2001, pp. 371–373.

5

SYSTEM TECHNICAL ASPECTS

5.1 INTRODUCTION

It was already shown that suggested structures of strip lines and meander lines as effective EM polarizers are suitable as the data encoding elements of the spatial-based system. The salient attributes of the polarizers were also discussed. Reaching to millimeter image resolution at a reading range of less than 50 cm mandates the 60 GHz band and above. This chapter looks at this new industrial, scientific, and mathematical (ISM) band and explores its technical and regulatory limitations.

In the second part of this chapter, the reader antenna is described. First, the technical and operational necessities of the reader antenna are discussed. Then, the available or proposed antennas at the 60 GHz band are reviewed to confirm that "off-the-shelf" products are not suitable for the demanded reader antenna specifications in the proposed approach. Finally, the suitable antenna candidate is suggested along with its design and measurement procedure.

5.2 The mm-BAND OF 60 GHz

5.2.1 Highest Attenuation, Lowest Regulatory Limitations

The frequency band 60 GHz is a bold region in the frequency spectrum as it has one of the highest attenuations due to the oxygen absorption. While 0.1–0.5 dB/km loss occurs from oxygen absorption in the frequency spectrum below 100 GHz, the region around 60 GHz experiences 15–20 dB/km loss [1] (Figure 5.1). Moreover to this unusual loss, the high frequency nature also suggests a higher free space attenua-

Advanced Chipless RFID: MIMO-Based Imaging at 60 GHz–ML Detection, First Edition.
Nemai Chandra Karmakar, Mohammad Zomorrodi, and Chamath Divarathne.
© 2016 John Wiley & Sons, Inc. Published 2016 by John Wiley & Sons, Inc.

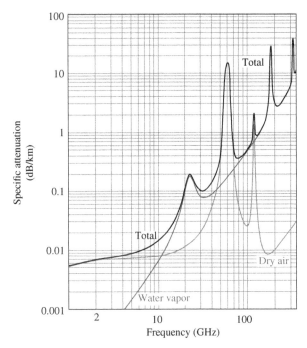

Figure 5.1 Specific attenuation due to atmospheric gases [1].

tion of 128 dB/km based on Friis transmission equation. Therefore, the combination of these two high attenuations suggests the band 60 GHz band as an unsuitable candidate for long-range communication services. This is the main reason that this band is generously released for ISM usage by regulators as it has no benefit for other high-revenue commercial services. Although the high attenuation may be viewed as disadvantageous, it provides the possibility of frequency usage of the band in proximity without fear of interference, which can be seen beneficial for certain applications. For instance, intersatellite links at 60 GHz are well protected from any terrestrial interference by high atmospheric attenuation. The 60 GHz band has also been an attractive frequency candidate for military and highly secure communication services regarding its severe absorption by building materials, which could ensure that signals cannot travel far beyond their intended recipient. Thankful to rich legacy of these applications, a wide variety of components and subassemblies for 60 GHz products are available today for ISM band users [2]. The above-mentioned characteristics of the 60 GHz frequency band have stimulated the regulators of telecommunications to recognize this frequency band as an unlicensed option for short-range indoor/outdoor applications [3]. According to the Radio Regulations (RR) prepared by the International Telecommunication Union (ITU), the bands 55.78–66 GHz are allocated to fixed and mobile services in all three ITU regions. Also, footnote 5.547 of the ITU Radio Regulations designates the bands 55.78–59 GHz and 64–66 GHz

TABLE 5.1   Available Spectrum on 60 GHz Band

Country/Region	Frequency (GHz)
Australia	59.4–62.9
United States, Canada, and Korea	57–64
Japan	59–66
Europe	57–66
China	59–64
$EIRP_{max}$	40 dBm
P_{trans}	10 dBm
$Bandwidth_{min}$	100 MHz

as a suitable candidate for high-density applications in the fixed service [3]. Following the ITU allocation, most countries have recognized some specific portions of 55–66 GHz as ISM frequency bands [4,5]. Table 5.1 shows the released band by some regulators.

Moreover, the maximum transmitting power limitations for interference reduction or human exposure issues for this frequency range are less limited by regulators. The maximum EIRP of 40 dBm and maximum power to the antenna of 10 dBm are normal values that cover the regulatory limitations in most countries. The minimum frequency bandwidth is also set to 100 MHz, and therefore, most commercial producers have selected it as their channel bandwidth [4,6–8]. Hence, it can be concluded that the technical and regulatory situation of the 60 GHz frequency band makes this frequency range an ideal candidate for imaging purposes in RFID applications.

5.2.2   Lowest Tag Printing Cost

It is well known that optimal tag reflection corresponds to metallic tags that are at least as thick as the skin depth. The skin depth as a function of radio frequency and material conductivity is shown in Figure 5.2.

Near-field-communication (NFC) tags at 13.6 MHz typically use >20 μm stamped aluminum, whereas UHF antennas (960 MHz) need only 5–10 μm stamped aluminum or plated copper. Although the cost of stamped aluminum at NFC or ultrahigh frequency (UHF) systems is negligible compared to the cost of chip in each RFID tag, aluminum cost plays a major role in determining the tag cost for the fully printable chipless RFID tags. On the millimeter-wave band 60 GHz, the required thickness reduces to below 1 μm, which can be interpreted as an extremely low-cost tag structure even cheaper than barcodes. This process also insures low AC skin losses due to the smooth planar metal film surface versus rough metal surface that can have higher losses [9]. Other methods for printing RFID tag include nanosilver ink printing with postflash cure to solidify metal pigments or thermal film metal transfer printing with postplating. However, each of these processes requires additional steps to build up the required optimal metal thickness. This is true, for example, of the Kurz SECOBO® thermal transfer antennas shown in Figure 5.3. However, as the RFID frequency of

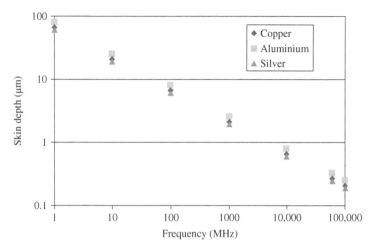

Figure 5.2   Tag frequency versus skin depth in microns for various materials [9].

Figure 5.3   Kurz SECOBO® antennas [10].

operation continues to increase, the required skin depth thickness decreases. Thus, from the perspective of the metal film thicknesses necessary for efficient interrogation signal reflection, there are tremendous advantages in utilizing higher frequencies for chipless RFID operation. Thinner metal RFID antennas mean more viable lower cost tag fabrication techniques such as inkjet and metal foil transfer can be used with a higher throughput. However, surface roughness also becomes more important at higher frequencies as high current densities occurs near the metal surface.

Apart from the required metal thickness for chipless RFID tags at the 60 GHz band, the millimeter-band of operation means miniaturized EM polarizers. This has two advantages. First, the tag cost also reduces as the lower amount of metal is needed to shape a tiny tag. Second, the content capacity of the tag per unit area significantly enhances compared with other proposed techniques in lower bands.

5.3 READER ANTENNA

5.3.1 Technical and Operational Requirements

In this section, it is intended to derive all the technical and operational specifications of the reader antenna in the proposed spatial-based technique. Based on these necessities, the antenna type and its characterizations are later constructed.

As the system is based on the separation of transmit and receive signals through an orthogonal polarization scheme, the ability of the antenna to distinguish two polarizations is very important. This characteristic of the antenna is known as its cross-polar level (*CPL*) that defines the quality of the signal polarization. In other words, the ratio of the desired polarization to the orthogonal/cross-component is known as the antenna *CPL*:

$$CPL = \frac{\text{Desired field component}}{\text{Cross-polar field component}}, \quad 1 \leq CPL \leq \infty;$$
$$CPL(\text{dB}) = 20 \times \log(CPL) \tag{5.1}$$

In the proposed system, both transmit and receive antennas are close to each other; therefore, the high *CPL* in the order of 15–20 dB is essential for the proper function of the system. Otherwise, the direct signal leakage from the transmitter to receiver due to low isolation of the antennas may saturate the receiver and affect the tag reading process. The higher the *CPL*, the better is the receiver sensitivity with respect to the cross-polar tag response.

As discussed before, for providing fine azimuth resolution, the reader moves around the tag and illuminates it through different view angles. This mandates a wide radiation pattern of the antenna in azimuth. The length of the route on which the reader will move are known as the synthetic aperture length depends on the tag size and the reading distance. However, for the proposed application this length is around 15–30 cm at 10 cm reading distance. Figure 5.4 shows the reading distance and the required half-power beamwidth (*HPBW*) of the antenna in the azimuth direction. The synthetic aperture length of 30 cm is considered, as the maximum length, in Figure 5.4. The angle θ is therefore

$$\theta = \tan^{-1}\left(\frac{10}{15}\right) = 33^\circ \tag{5.2}$$

And from Equation (5.2), the *HPBW* would be

$$HPBW = 2 \times (90^\circ - \theta) = 114^\circ \tag{5.3}$$

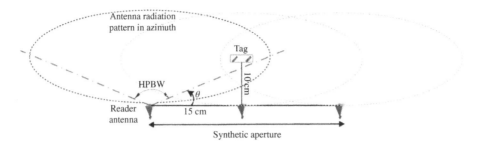

Figure 5.4   Required *HPBW* of the antenna.

The *HPBW* requirement of 115° is then selected as the worst case for the reader antennas.

The radiation pattern of the antenna in the elevation angle is not required to satisfy any specific limitations; however, it is desirable if the antenna focuses energy only on the tag surface. This minimizes the multipath issue and reflections by other objects around the tag. Although it was already justified that the proposed system is very robust to multipath interferences, the limited pattern in elevation minimizes the energy wasted in unwanted directions.

Considering the requirements for physical movement of the antenna, the size of the reader antenna shall be kept small and lightweight for ease of movement for handheld applications. Therefore, bulky antennas are not of interest of this application. A fine-range resolution of millimeter order was linked to a huge bandwidth requirement that is impractical. However, for minimizing the unwanted reflection and also lower exposition of energy toward unwanted directions, it is beneficial to have as fine range resolution as possible. Therefore, it is desirable that the antenna would be able to cover the whole ISM frequency band of 57–64 GHz. This implies around 12% bandwidth requirements for the antenna.

In conclusion, it appears that the followings are the intended requirements of the reader antenna(s) for the proposed spatial-based chipless RFID system:

(i)   Operating at frequency band 57–64 GHz

(ii)   High *CPL* in boresight direction, *CPL* > 15 dB

(iii)   Fan-shaped radiation pattern; wide azimuth beamwidth >115° azimuth and limited elevation beamwidth <15°

(iv)   Lightweight and small size for handheld applications of the reader.

5.3.2   Off-the-Shelf Products

It was mentioned that the 60 GHz band has been already used for intersatellite communication links and for highly secure terrestrial mobile communication. The recent release of 57–64 GHz as an ISM band accelerates many products on this band for public usage [4,5,11,12].

However, the higher attenuation of signals related to the 60 GHz band suggests a high-gain antenna for short-range communication links, in the order of few kilometers. There are some antennas available on the market that cover the intended frequency band; however, their radiation pattern and impedance bandwidth appear to be unsuitable for the purposed chipless RFID system. Nokia's radio link Metro Hopper [13] is a high-data-rate communication link but only covers 57.2–58.2 GHz and has a large antenna size that is not suitable for the proposed application. Proxim [14] suggests products at 60 GHz that utilize antennas for covering the whole band of 57–64 GHz. The gains of antennas range from 25 to 35 dB$_i$. There is no information about the radiation pattern but high antenna gain suggests a large aperture size and narrow radiation pattern, which do not satisfy the required specifications for the spatial-based system. Moreover, no information about the cross polarization level is disclosed. As the ISM band of 57–64 GHz has been taken by companies for providing high-data-rate communication in point-to-point links [2,3]; therefore, it is expected that the commercially available antennas have high gain and narrow radiation patterns. Moreover, the antenna size is not an issue in those applications. Therefore, it is unlikely to be able to find an off-the-shelf antenna that satisfies the necessities of the reader antenna for the proposed chipless RFID application.

5.3.3 Array of Printed Dipoles

The general specifications of the required antenna outlined in the previous section govern the type of suitable solution for the proposed reader antenna. One may suggest that an array of printed dipole antennas is a suitable candidate to meet the specification requirements. The printed antenna adequately satisfies the requirements of low profile, lightweight, and low-cost attributes of the reader system. Dipole antennas provide wide radiation patterns in the azimuth direction, while their *HPBW* in the elevation angle depends on the dipole length. Moreover, the array configuration provides the opportunity for controlling the beamwidth in the desired direction, here the elevation angle. Dipoles are normally linearly polarized antennas and are able to provide adequate *CPL* for many applications. The compact size of the dipoles is another positive aspect that is beneficial for the proposed system. Compared with the microstrip patch antenna, the printed dipole enjoys a few advantages. They provide wider bandwidth, lower loss, and less parasitic radiation of feedlines [15–18]. Therefore, the candidate reader antenna addressing the reader's technical and operational specifications is a double-side printed dipole (DSPD). Compared with a single-side printed dipole, the DSPD yields broader bandwidth and higher gain [19,20].

Nesic *et al.* [15] reports a one-dimensional array of printed dipoles fed by a microstrip line at 5 GHz with 7% bandwidth. The antenna works on its third resonance so provides a higher gain than that of its first resonance. However, the very low antenna bandwidth on its third resonance has been mitigated by penetrating the feedline between two slots. No information on the radiation pattern of the antenna was disclosed. Scott [20] showed a microstrip-fed printed dipole array using a microstrip-to-coplanar strip line (CPS) balun. In Refs [15,20], the balun was not easy to match the impedance and the structures were too big and bulky; hence, it was complicated to build an array structure [21].

Brankovic and Nesic [22] patented a pentagonal dipole antenna on a 64-element array configuration yielding 25% bandwidth at 60 GHz. The radiation pattern of this work is not wide in the azimuth direction and also no information of the *CPL* was declared. Considering the pentagonal radiator shape, the lower *CPL* is expected than that of a normal dipole. Suh and Chang [21] reported a 30 GHz printed dipole that has microstrip-fed CPS Tee junctions and yields 3% bandwidth. Alhalabi and Rebeiz [23] showed a high-efficiency array of printed dipoles designed at 22–26 GHz. Varying feedline lengths create prescribed phase shifts between the array elements. The same authors also proposed a 60 GHz single dipole antenna [24] suitable for laptops and portable communication devices. From the above-mentioned reviews, it can be concluded that none of the reported antennas at 60 GHz are suitable for the proposed RFID system. However, the approaches proposed in Ref. [23] are generally followed to build a suitable reader antenna for the proposed application. As discussed before, the reader antenna for the proposed application requires satisfying a set of specifications that are necessary for different reasons. An array of DSPD fulfils all of the requirements outlined previously and repeated in Table 5.2.

In the design process, the first step is to select the appropriate substrate for the printed dipole antennas. The suggested dielectric is Taconic TLX-8 since the author's group has an extensive experience with its performance at 24–27 GHz [25]. However, as the constitutive parameters, such as dielectric constant ε_r and loss tangents tan δ of Taconic substrate TLX-8, are not known at 60 GHz, a special test jig is built to measure the substrate parameters at 60 GHz. The next step is to design the dipole as the array radiating element. The dipoles are equipped with an unbalanced feedline, which is then converted to a microstrip line. A microstrip to coplanar waveguide (CPW) feedline transition facilitates a V-type input connector assembly. The array structure is precisely optimized to cover the entire band of 57–64 GHz yielding 12% bandwidth at 60 GHz. The flowchart of the DSPD array antenna design is shown in Figure 5.5. Each section of the design process outlined in Figure 5.5 is presented

TABLE 5.2   Technical and Operational Requirements of the Reader Antenna

Center frequency	60.5 GHz
Bandwidth	12%
Radiation pattern	Maximum 15° on E-plane Minimum 115° on H-plane
Cross-polar level	>15 dB
Substrate	✓ Low thickness for high cross-polar level
	✓ Stable electrical and mechanical performances on millimeter-wave region
Gain	Between 5 and 10 dB$_i$
Dimension	As small as possible (centimeter order)

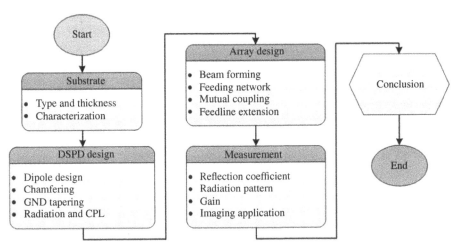

Figure 5.5 Design flowchart for array of DSPDs.

separately, and then the fabricated array is tested proving its actual potential for the proposed application.

5.3.4 Substrate Selection and Characterization

5.3.4.1 Substrate Selection and Thickness The first step of the antenna design is the selection of substrate. The design frequency and operational bandwidth of the antenna govern the selection of substrate. The thickness, relative permittivity, loss tangent, conductive strip thickness, and finally, mechanical and thermal stabilities are the critical parameters of the substrate at 60 GHz frequency band. Considering the 5 mm wavelength at 60 GHz, precision in the design and substrate properties plays critical role in the design success. Taconic TLX-8 would be a suitable candidate as it satisfies many of the aforementioned parameters [26,27]. Its low dielectric constant enhances the antenna bandwidth [18] and its favorable thickness is required for the design.

The selection of substrate thickness is bounded by two factors. A thin substrate is required for an efficient radiator. However, the large operational bandwidth needs a thick substrate. The substrate thickness should satisfy Equation (5.4) to ensure the suppression of higher order radiation modes on the dielectric surface [28]. At 60 GHz, the thickness of the substrate must be less than 1.06 mm.

$$\text{Thickness} < \frac{\lambda_0}{4 \times \sqrt{\varepsilon_{r\ eff} - 1}} \approx 1.06\,\text{mm} \tag{5.4}$$

In Equation (5.4), λ_0 is the wavelength and is between 4.68 and 5.26 mm. $\varepsilon_{r\ eff}$ is the effective permittivity of the substrate. Considering the available commercial substrate, Taconic TLX-8 with 0.127 mm thickness is selected. This thickness adequately satisfies Equation (5.4) and ensures that only one surface wave, TM_0 with zero cutoff frequency, will exist in the substrate [29]. This thin substrate also enhances the *CPL* of the antenna [30], which is the critical aspect of the design.

5.3.4.2 Substrate Characterization   The knowledge of actual electrical parameters of the substrate is vital for accurate design, especially in the millimeter-wave range, where dimensions based on electrical characteristics such as substrate permittivity and loss tangent are tolerance-critical. Today, most substrate manufacturers specify the dielectric parameters of their materials at 10 GHz. Consequently, in realized designs at millimeter-wave, frequency shifts attributed to permittivity deviation commonly occur [31]. Microwave measurement techniques to determine substrate permittivity and loss tangent at specific frequencies are (i) transmission and reflection and (ii) resonance techniques. Transmission and reflection techniques can be used to obtain the real part of permittivity for a high- or low-loss specimen but lack the resolution to measure low loss tangents. Resonant technique is only suitable for low-loss materials since the resonance curve broadens as the loss increases [32]. Usually, dielectric property measurement by resonance technique provides higher accuracy than that obtained from transmission and reflection technique. It is therefore considered as the most accurate method for dielectric constant and loss tangent measurement of planar structures. Considering the structural dimensions at millimeter-wave, the substrate integrated waveguide (SIW) resonators approach is proposed for substrate characterization at V- and W-bands [33].

Taconic (www.taconic.com, accessed on May 6, 2014) published data sheet of TLX-8 at 10 GHz reveals the following information: $\varepsilon_r = 2.55$ and $\tan \delta = 0.0019$. In this work, SIW cavity resonators are designed and fabricated at 60 GHz (Figure 5.6). Propagation and attenuation constants of an SIW resonator are related to modal expansion theory of a rectangle waveguide [34]. Equation (5.4) shows the equivalent dimensions of the rectangle waveguide on which a, b, d, and s are defined at Figure 5.6(a). The photograph of a prototype SIW resonator is shown in

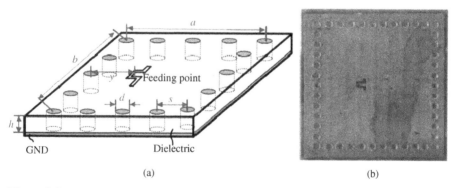

(a)   (b)

Figure 5.6   (a) Layout of a sample SIW resonator with via holes [24] and (b) photograph of prototype SIW resonator [33].

Figure 5.6(b). The resonance frequency of the rectangle SIW resonator for TE_{mn0} mode relates to Equation (5.6).

$$x_{eff} = x - 1.08\frac{d^2}{s} + 0.1\frac{d^2}{x}, \quad x = a, b \tag{5.5}$$

$$f_{mn0} = \frac{c}{2\pi\sqrt{\varepsilon_r}}\sqrt{\left(\frac{m\pi}{a_{eff}}\right) + \left(\frac{n\pi}{b_{eff}}\right)} \tag{5.6}$$

In Equation (5.6), c is the speed of light in free space, a_{eff} and b_{eff} are found from Equation (5.5), and ε_r is the effective dielectric constant. To find the effective dielectric constant, the proposed approach by Zelenchuk *et al.* [33] is followed. Six prototypes of SIW resonators are fabricated and tested to measure the substrate parameters at 60 GHz (c.f. Figure 5.6). The average dielectric constant and loss tangent are measured to be 2.25 and 0.01, respectively, which are substantially different than those for published datasheet at 10 GHz. Table 5.3 summarizes the results. The measured data of the substrate are considered in designing the proposed DSPD array antenna.

5.3.5 Array Radiator Design

5.3.5.1 Dipole Design The DSPD element with its design parameters is shown in Figure 5.7. The length (L_a) of the dipole arm is a function of the thickness and dielectric constant of the substrate and mainly controls the resonance frequency of

TABLE 5.3 Summary of Dielectric Parameters of TLX-8

60 GHz	$\varepsilon_r = 2.25$	$\tan\delta = 0.01$
10 GHz	$\varepsilon_r = 2.55$	$\tan\delta = 0.0019$

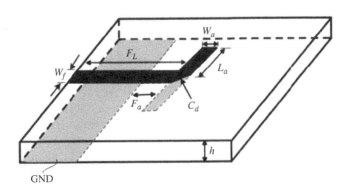

Figure 5.7 Printed dipole's structure and its dimensions, $W_f = 0.37$, $F_L = 1.05$, $F_a = 0.55$, $W_a = 0.19$, $L_a = 1$, $C_d = 0.2$ and $h = 0.127$ (all dimensions in millimeters).

the dipole. As the substrate does not cover the whole structure, the resonant length is not scaled with $1/\sqrt{\varepsilon_{r\,eff}}$ as an antenna in a homogeneous medium [29].

The initial value of the dipole length comes from the following formula [35]:

$$\text{Dipole length} = 0.47\nu/f, \quad \text{where } \nu = c/\sqrt{\varepsilon_{r\,eff}} \tag{5.7}$$

$$\varepsilon_{r\,eff} = \frac{\varepsilon_r + 1}{2} + \frac{\varepsilon_r - 1}{2}\left[\sqrt{1 + \frac{12h}{w_a}} + 0.04\left(1 - \frac{w_a}{h}\right)^2\right] \tag{5.8}$$

In these formulas, ν is the velocity of propagation inside the substrate and the rest of parameters are defined in Figure 5.7. Based on Equations (5.7) and (5.8), the required length of dipole arm is 1.2 mm. This initial length is entered in full-wave EM solver CST Microwave Studio and the design is optimized to meet the specification requirement. The simulated dipole arm length is 1 mm. The dipole arm width is independently optimized to match with 50 Ω input impedance although the optimized value of W_a would be different than that for the parallel feedline width, W_f. Matching of these two strip lines with different widths is manipulated through changing the chamfering depth, C_d. Finally, the parallel suspended feedline of the dipole arm is transformed to a microstrip line that is connected to a two-way power divider of the beam-forming network of the four-element array. To enhance the bandwidth and performance of the antenna, chamfering of the dipole element and tapered ground plane at the suspended stripline to microstripline transition are used. This technique yields more than 12% bandwidth at 60 GHz, while conventional design yields 7% bandwidth [20].

5.3.5.2 Dipole Chamfering   Dipole antennas are inherently unbalanced circuits and create distortion in radiation pattern. Connecting dipole to a balanced feedline requires a balun network to match balanced and unbalanced circuits. Chen and Sun [36] suggested a double-sided feed network on the substrate as the balun network for the planar dipole antennas. As shown in Figure 5.7, the double-side structure feedlines provide a smooth transition between the unbalanced and balanced currents of the line and the dipole feed. This configuration also simplifies the balun network. As discussed before, the chamfering of the dipole edge is required to match two different widths of the feedline and dipole arm. Moreover, the right angle in the microstrip line creates extra capacitance and also changes the electrical length of the line [37]. These two will affect the impedance matching of the structure that should be compensated by chamfering. For symmetry, both dipole arms at their right angle transitions are chamfered. However, from our detailed investigation in full-wave CST Microwave Studio simulation, we found that only top arm chamfering provides best results. Figure 5.8 shows the input impedance return loss (S_{11} in dB) versus frequency for three cases – top arm, both arms, and no chamfering. Considering the effect of different chamfering scenarios on the input impedance, the maximum RL of 27 dB is obtained with top arm chamfering. As the bottom arm is directly connected to the ground, then it probably has minor effect on the impedance matching and only top arm chamfering provides the optimum impedance matching.

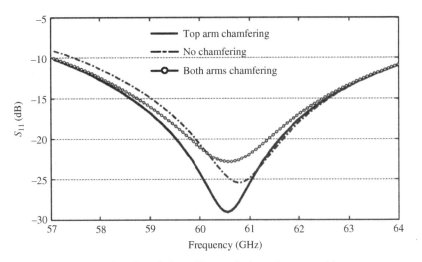

Figure 5.8 Chamfering effect on the impedance matching.

5.3.5.3 Ground Connection Tapering The next investigation to improve the performance of the dipole is a tapered ground connection of the bottom feedline of the balanced feed network of the DSPD antenna. The ground plane acts as a reflector to the dipole, hence largely influences both the radiation pattern and input impedance. Therefore, the shape and distance of the ground plane from the dipole element determines the input impedance bandwidth, beamwidth, gain, and cross-polar level of the dipole element. The authors have investigated the effect of the ground plane on monopole and dipole elements' performance in Refs [38,39]. We have investigated two types of ground planes – conventional (also called abrupt) and tapered ground planes. The effect of the tapered ground connection depends on its tapering angle. The tapering angle (α) is shown in Figure 5.9. The angle α is changed to find the best performance of the dipole through parameter sweeping. A tapered connection causes a resonance shift in return loss figure. To accurately compare the two scenarios, few parameters of the top arm are retuned to have the same frequency resonances. Figure 5.9 shows the input impedance return loss versus frequency for both abrupt and tapered connections. It confirms that the abrupt connection proposes better impedance matching on the resonance frequency while not affecting the bandwidth of the radiator. This effect was also experienced over the lower frequency range of 20–26 GHz [23].

5.3.5.4 Dipole Radiation Pattern and Cross-Polar Level The radiation pattern of a single DSPD is simulated with full-wave EM solver CST Microwave Studio and the result is shown in Figure 5.10. The 2D radiation patterns of a DSPD are depicted for three sample frequencies – 58, 60, and 63 GHz in Figure 5.11. As will be discussed later, the final array configuration is a linear array while elements are oriented vertically. Through general knowledge of the array theory, one can conclude

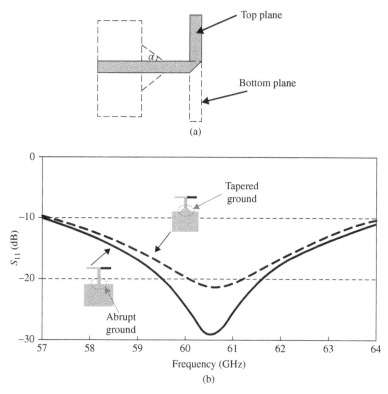

Figure 5.9 (a) Embedded dipole with tapered GND and (b) effect on impedance matching.

that the H-plane of the final array structure is similar to the single DSPD, while its E-plane would be different from that of a single DSPD.

The 3-dB beamwidth of the single DSPD on its H-plane, for three sample frequencies, is shown in Table 5.4. The values of the 3-dB beamwidth adequately satisfy the wide radiation pattern requirements of the final array as discussed in Table 5.2. This wide radiation pattern ensures that in the synthetic aperture approach, the tag's surface receives enough energy from the reader at different view angles. As is evident from the H-plane radiation pattern shown in Figure 5.11, the maximum radiation does not occur on the broadside direction ($\varphi = 0$) as expected. Instead, the pattern of a single DSPD is shifted to the left direction. This is because of existence of ground plan in one side of the substrate that pushes the radiation in the opposite direction. The direction of maximum radiation depends on the frequency and is approximately shown by the black dashed line in Figure 5.11; however, it varies from $-345°$ to $-350°$.

As discussed earlier, the quality of the antenna polarization known as its *CPL* is a critical parameter in the proposed application. For the designed single DSPD, the *CPL* is simulated by having co- and cross-polar probes in CST Microwave studio and measuring the S_{21} of the link. The simulated *CPL* for a single DSPD is

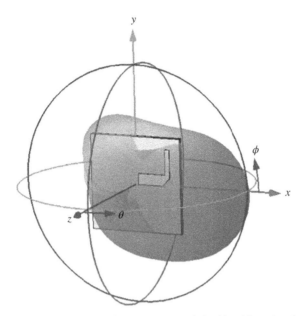

Figure 5.10 CST-generated 3D radiation pattern of double-side printed dipole (DSPD) antenna.

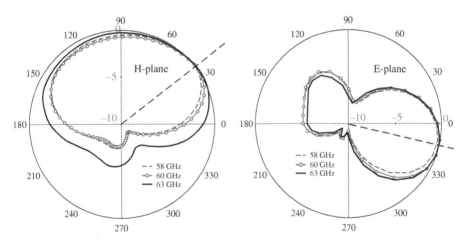

Figure 5.11 Simulated radiation pattern (E- and H-planes) of DSPD.

depicted in Figure 5.12. Over the entire frequency range of operation, this ratio for a single dipole is less than −23 dB. There is the opportunity of decreasing the *CPL* by simply using the defected ground structure (DGS) [29]. However, the provided −23 dB *CPL* through the DSPD adequately satisfies the requirements of the proposed application.

TABLE 5.4   3-dB Beamwidth of a Single DSPD in H-Plane

Frequency (GHz)	3-dB Beamwidth (°)
58	194.8
60	197.1
63	201

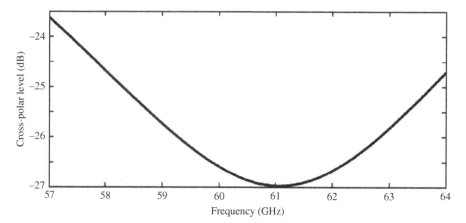

Figure 5.12   Simulated cross-polar level of single dipole (DSPD).

5.3.6   Array Design

5.3.6.1   Required Beam-Forming Array   The antenna specifications discussed earlier govern the array structure. As a fan-shaped radiation pattern, wide azimuth and limited elevation, is required hence a linear array-type antenna would satisfy this specification. Multiple vertically aligned dipoles restrict the radiation pattern in the elevation angle. It was calculated that four dipoles are enough to satisfy the 15–20° required beamwidth in the elevation angle. The wide H-plane radiation pattern of a single dipole as shown in Figure 5.11 also fulfils the 115° required azimuth beamwidth. To select a proper feeding network, the overall system structure for the spatial-based chipless RFID application will be considered. As the SAR-based signal processing is based on the relative position of the reader and tag, it is mandatory that the antenna has a unique and fixed radiation pattern in all positions. This is the reason why a phased array antenna type would not be suitable for this application as it creates various beam shapes on different angles [40]. From the array theory, it is clear that the uniform excitation of the array's elements creates the maximum radiation intensity and minimum 3-dB beamwidth in the boreside direction [41]. The requirement for a fixed fan-shaped radiation pattern suggests a linear array of dipoles with uniform excitation.

5.3.6.2 Corporate Feeding Network

5.3.6.2 Corporate Feeding Network The requirement for a uniform excitation of the array elements and also having two separate Tx and Rx antennas at the reader side ease the power divider arrangement. A T-junction power divider as one of the simplest type of power dividers [42] would be suitable for the proposed application. It is fully printable and of low cost without any lumped element in the structure. These specifications make the T-junction power divider very suitable for feeding the DSPDs on the array. However, the T-junction power divider suffers from two major limitations. First, it utilizes quarter wavelength transformer in its structure that restricts the operational bandwidth of the element. Second, it provides low isolation among output ports. The low isolation would be a major limitation for monostatic systems on which the same antenna is used for transmit and receive purposes. While in the proposed bistatic configuration, two separate antennas are utilized; therefore, the low isolation of the output is not an issue. The low bandwidth of the power divider, due to the quarter wavelength transformer, can be compensated by minor changes in the component design. Figure 5.13 shows a conventional two-way T-junction power divider and its dimensions. The length of the quarter wavelength transformer is fixed and it is intended to transform $50\,\Omega$ impedance to the $100\,\Omega$. The length of the $50\,\Omega$ arm on each side of the power divider depends on the required physical distance between two connected dipoles. The chamfering depth is optimized through parameter sweep in the CST simulation tools.

It is normally accepted that for providing better impedance matching, insertion of a small V-notch in the dividing point of the T-junction structure is a good technique. This gap is shown in Figure 5.14(a). In the current work, however, it was found that a reactive stub acts better and can enhance the power divider bandwidth.

The corporate feed network of the proposed array antenna includes three two-way T-junction power dividers as shown in Figure 5.15. The S-parameter of the corporate feed network is depicted in Figure 5.16 that shows that S_{11} in decibel versus frequency is adequately below $-15\,dB$ over the whole frequency range of operation. The minimum value of the reflection coefficient is $-20.5\,dB$ at $61.5\,GHz$. This is a noticeable result considering the narrowband aspect of the quarter wavelength sections in the power divider structure. Normally, four transmission coefficient curves are expected in Figure 5.16. However, as the transmission coefficients of each port (S_{j1} where $j = 1$, 2, 3, 4) are similar, only two of them are depicted as an example.

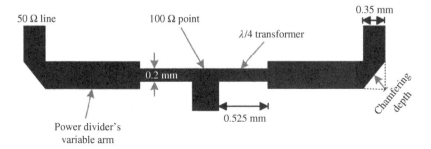

Figure 5.13 T-junction power divider.

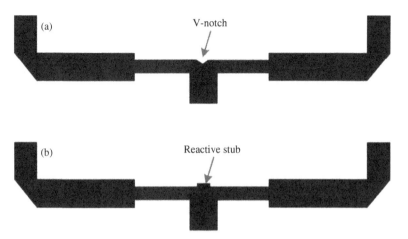

Figure 5.14   T-junction power divider with (a) V-notch and (b) reactive stub.

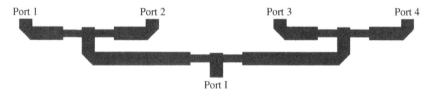

Figure 5.15   Complete corporate feeding network of array.

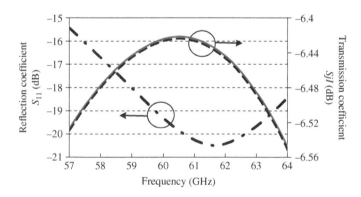

Figure 5.16   The S-parameter of the corporate feeding network of array.

The values of the transmission coefficient vary between 6.4 and 6.55 dB over the band. While the reflection is adequately low, almost a quarter of the input power is delivered to each port. Therefore, the array elements are uniformly excited. Figure 5.17 shows the changes of the transmission coefficient phase on four different output ports. The phases of ports 1 and 2 are exactly equal. The same is also valid

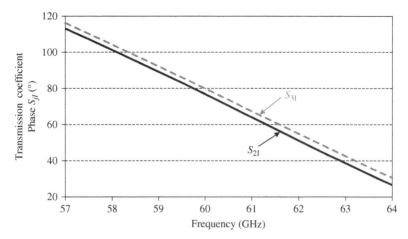

Figure 5.17 Transmission phase of different output ports (S_{21} and S_{31}) versus frequency.

for the phases of ports 3 and 4 in Figure 5.15. There is a small difference between the phases of ports 1 and 2, and 3 and 4 as shown in Figure 5.17. As the difference is very small, 2–3°, it is not a serious issue in the total radiated power.

5.3.6.3 *Mutual Coupling*   Mutual coupling between the array's elements is a critical factor in the array design that can negatively affect the total performance of the array. Increasing the interelement distance reduces the mutual coupling; however, a large distance between dipoles results in grating lobes in the radiation pattern if the inner-element distance exceeds the wavelength, λ.

While the half-wavelength distance is normally selected as the inner-element space, larger distances ($\lambda/2 < d < \lambda$) are also feasible if narrower *HPBW* is desirable. In the proposed application, smaller *HPBW* on the elevation angle is desired as it focuses more energy on the tag surface and minimizes the reflections from surrounding objects. Therefore, the power divider's arm is tuned to provide 3.45 mm inner-element spacing equivalent to 0.65–0.73λ over the band (Figure 5.18). This large distance between the elements reduces the *HPBW*, as well as minimizes the mutual coupling.

To simulate the isolation between two adjacent dipoles, two ports are connected to the dipoles and the transmission coefficient, S_{21}, is simulated in CST. Figure 5.18 shows the simulated result. The mutual coupling between two adjacent dipoles is adequately below −22 dB. It is important to verify that the low value of S_{21} is adequately showing the mutual coupling and it is not due to the high loss occurred in the feedline of the dipoles. This simply means that if the feedline causes a noticeable amount of loss, then a good S_{21} is resulted. Hence, loss occurring on the feedline can be seen as an error on the transmission coefficient measurement. This feedline, which may degrade the S_{21} result, is highlighted in Figure 5.18. The feedline length is only 0.8 mm (0.16λ) and therefore the occurred loss is very minor. By using the CST

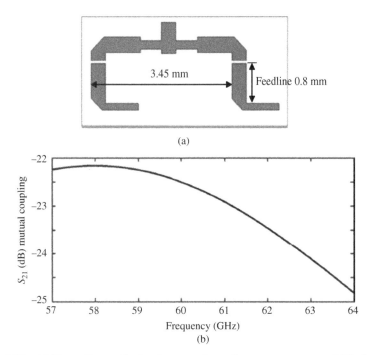

(a)

(b)

Figure 5.18 (a) Two adjacent dipoles for mutual coupling measurement and (b) simulation result.

and/or advanced design system (ADS) facilities, it can easily be seen that the feedline loss only contributes 0.17–0.23 dB excess loss to the result, S_{21} of Figure 5.18. This means that the isolation between adjacent elements is below -22 dB over the whole frequency band of operation.

5.3.6.4 *Extended Feedline* The designed array of four dipoles and its feeding network has acceptable electrical performance over the frequency range of operation. However, four-element dipole array is too small that is impractical to connect the array with a V-type end launch connector as shown in Figure 5.19. A V-type connector that is suitable for the frequency range of operation has a much larger size than the designed antenna. To address this issue, the array's feedline is extended for $4\lambda \approx 20$ mm to provide (i) enough space for connector installation and (ii) minimize the cross-polar component.

Providing enough space for connector installation, it is possible to simply extend the microstrip line; however, induced surface wave results in higher feedline loss. To reduce the effect of long feedline on the antenna performance, the CPW extension is used instead of microstrip line. The grounded CPW line is immune to surface wave and also it provides better resistivity to the copper roughness than the normal microstrip line [43]. The prototype array is fabricated on TLX-8 using photolithographic fabrication process. The photograph of the assembled array antenna with

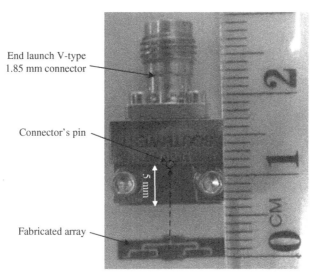

Figure 5.19 Comparing physical sizes of the array antenna and V-type connector.

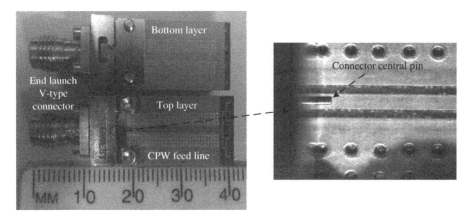

Figure 5.20 Extended CPW feedline and final array structure.

the extended CPW feedline and the V-type connector [44] assembly is shown in Figure 5.20.

As it is clear from Figure 5.20, the length of the extended feedline seems to be longer than what is needed for the purpose of V-type connection. The feedline length is bound by two contradictory factors. First, the shorter line is preferred to minimize the loss occurring on the line and avoid unwanted radiation from the feedline. This implies that a 5–7 mm extension is enough. The increased *CPL* due to the back-lobe reflection by the large connector body is another affecting parameter. As shown in Figure 5.11, the dipoles have a noticeable back-lobe. If the array antenna is placed very close to the connector, its back-lobe energy is reflected back on the broadside direction by the connector structure. This may disturb the radiation pattern of the array antenna. However, the major issue would be related to the increased level of the array *CPL*. To minimize this effect, the length of the feedline is extended beyond what was mandated by the physical requirements for connection purposes. As a trade-off selection for the feedline length, 20 mm extension is selected for the length of the feedline. This provides enough distance between the connector and the array for minimum disturbance of the radiation pattern and degradation of the *CPL*. However, the loss occurring on the line will also be considered.

5.3.7 Measurement Results

The prototype four-element dipole array antenna shown in Figure 5.20 is first characterized for its input impedance return loss. This evaluates the general performance of the structure as an effective antenna at the desired frequency band of 57–64 GHz. Then, the radiation patterns of the array in both E- and H-planes are measured. This verifies that the radiation pattern of the array is suitable for the SAR signal processing purposes. The *CPL* of the radiation pattern in both E- and H-planes are also considered and calculated to ensure that the array is adequately capable of picking up the intended polarization while cancelling the orthogonal polarization component. Finally, the average gain of the antenna is measured over the entire frequency band. In all measurement steps, full two-port error correction calibration and confidence check of the Agilent performance network analyzer (PNA) E8361A are performed to ensure accurate results.

5.3.7.1 Reflection Coefficient Measurement   The designed array of four DSPDs is simulated using CST Microwave Studio. However, the V-type connector may have significant effect on the final performance of the structure due to its large size compared with array antenna. Therefore, the connector is also included in the simulation process as shown in Figure 5.21. The dimensions of the V-type connector are derived from its data sheet [45]; however, aluminum (lossy metal) is assumed for the connector body and silver for its central pin. The fabricated antenna through the photolithographic process shown in Figure 5.20 is also connected to the Agilent PNA for its return loss measurement after proper system calibration.

The simulated and the measured S_{11} in decibel versus frequency are presented in Figure 5.21. It is important to reemphasize that the simulated and measured return loss reflects the effects of the array antenna, its long feedline, microstrip-to-CPW transition, and the V-type connector assembly. Three main aspects of the graphs in Figure 5.21 can be categorized as follows:

(i) *Satisfactory Return Loss.* As shown in Figure 5.21, both simulation and measured results yield more than 10 dB return loss over the frequency band of operation with maximum RL of more than 30 dB at 63 GHz. This ensures the satisfactory performance of the fabricated array over the band 57–64 GHz. It also confirms that the effects of a long feedline and the connector as well as

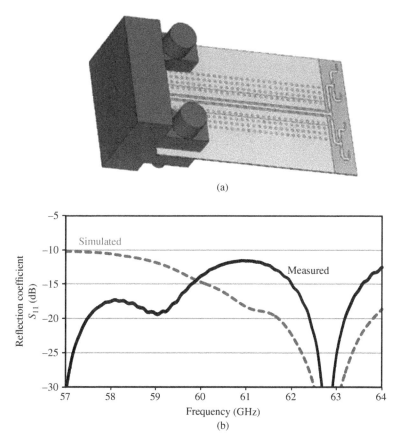

Figure 5.21   (a) Array and V-type connector for simulation and (b) reflection coefficient.

the transition between microstrip to CPW have been appropriately taken into account.

(ii) *Minor Simulation and Measurement Mismatch.* The graphs in Figure 5.21 are not fully matched to each other. Regarding the high frequency range of operation and also the fabrication process, some minor mismatching would be normal. To have a better understanding of the problem, the fabricated array is investigated thoroughly with a digital microscope. The dimensions of the array in different sections are measured and compared with the original values. The percentage of errors in various parts of the array is shown in Figure 5.22. As one can see from Figure 5.22, errors in different parts of the structure have significantly different values. While the width of the design is more vulnerable to errors, the lengths of the various parts are more immune to fabrication errors. For example, while some widths are experiencing 11% error, the maximum error in the array lengths is found to be only 0.4%. Inaccuracy in the line width significantly affects the impedance matching as it

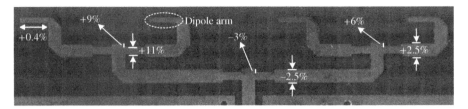

Figure 5.22   Percentage of errors in fabrication process.

changes the line impedance. The effect of errors in line lengths are discussed when the dielectric characterization is being considered. In addition to the errors occurring in the line widths, which result in impedance mismatch, one may suspect that residual surface waves, in the extended via grounded CPW line, and V-type connector assembly, cause impedance mismatch between the simulation and measured results. However, more than 10 dB return loss over the frequency band ensures no significant mismatch in the antenna and meets the specifications as discussed earlier.

(iii) *Accurate Dielectric Characterization.* The resonance frequency of the array is mainly controlled by the length of the dipole arm and the dielectric constant. As shown in Figure 5.21, the simulated resonance frequency and the measured resonance are both concentrated around 63 GHz. This means that the simulation and measured results are adequately matched from the resonance frequency perspective. As was already explored in (ii), the lengths of the dipole's arms are not significantly affected by the fabrication tolerance. Therefore, having the same simulated and measured resonance frequencies confirms the accuracy of the dielectric characterization method as described earlier. In other words, the calculated mean value of the measured ε_r for Taconic TLX-8 at 60 GHz provides a good prediction of the substrate performance at the 60 GHz frequency band.

5.3.7.2   Antenna Fabrication Error   At 60 GHz, the design would be very sensitive to any error during fabrication process. The effect of fabrication error was shown in the previous section, where inaccurate widths in the array feed network and dipole arms resulted in mismatch between the measured and simulated reflection coefficients. At least two array antennas are required in the proposed technique; therefore, it is important to consider the design robustness to dimensional tolerances as process variation in photolithographic etching always exists. It was discussed earlier that the T-junction power divider provides higher bandwidths when a reactive stub is utilized compared to a conventional V-notch in its structure. Moreover, it was found that the reactive stub significantly influences the design robustness to process errors.

Based on this finding, five samples of fabricated array antennas as shown in Figure 5.23 are measured by a digital microscope. The values of errors in different sections of the array structure are summarized in Table 5.5. As discussed earlier, the

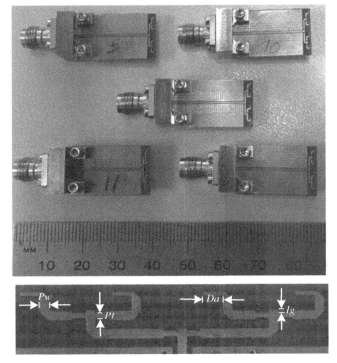

Figure 5.23   Five DSPD arrays and highlighted parametrized sections of array.

TABLE 5.5   Fabrication Variation

Parameter	P_w	D_a	P_t	I_g
Error (%)	$[-1.5, 3]$	$[-0.2, 0.4]$	$[-7, 11]$	$[-3, 9]$

width of lines in the printed circuit is more vulnerable to process error than the line length.

The measured input impedance, return loss (S_{11} in dB) versus frequency for the five fabricated array samples and the simulated S_{11} based on the accurate design values are shown in Figure 5.24. While the measured return losses are not completely matched with the simulation result due to the fabrication error, all five samples have acceptable performance ($S_{11} > 10\,\mathrm{dB}$) in the frequency range 57–64 GHz. Moreover, one may notice that all of the fabricated samples acquire almost the same resonance frequency as suggested through simulation, around 62.8–63.2 GHz. It can also be concluded that irrespective of fabrication error through conventional photolithographic printed circuit board (PCB) fabrication technique, appropriate antenna design at 60 GHz millimeter-band is feasible.

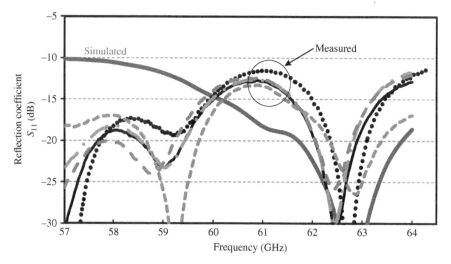

Figure 5.24 Simulated S_{11} and five measured return losses of sample arrays.

5.3.7.3 Radiation Pattern Measurement After satisfactory input impedance RL measurement, the radiation patterns of the fabricated array are measured. The array antenna under test (AUT) is set as a receiver antenna and a horn antenna [46], manufactured by Ainfoinc, is used as a standard gain transmitting antenna. Agilent PNA E8361A in its receiver mode is used. Figure 5.25 shows the photograph of the test setup with the standard gain horn antenna mounted on a fixed Perspex pole and the AUT mounted on a turn table. The radiation pattern is measured in every 10° steps for full 360° for both E- and H-plane cuts. Some conical absorbers are placed around the antennas to reduce surrounding interferences. Both measured and simulation copolar and cross-polar radiation patterns at 58, 60, and 63 GHz are shown in Figure 5.26. The measured radiation patterns are in congruence with the simulation results with minor shifts in sidelobe levels.

The measured cross-polar result is shown in Figure 5.26. As is clear from Figure 5.26, the *CPL* is adequately below −20 dB. This is a very satisfactory result considering the effects of the long feedline and the reflection from the large V-type connector.

As it is clear from Figure 5.26, the E-plane pattern is narrow compared with H-plane as the array is vertically aligned. Therefore, the 10° measurement would not be sufficient for the E-plane and to accurately measure the *HPBW* of the antenna in E-plane, measurements are repeated for every 1° for the ±10° of the boresight direction as shown in Figure 5.27. By considering the manual rotation of AUT in every 1°, small errors may be expected. To minimize this error, the measurement is repeated three times and the average value is considered. Figure 5.27 shows the copolar radiation pattern in ±10° of boresight direction of the array antenna at 58 GHz. It shows that the measured *HPBW* is smaller compared with the simulated *HPBW*. Table 5.6

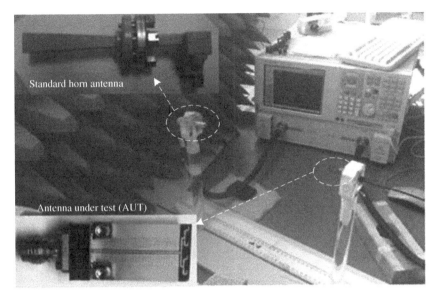

Figure 5.25 System setup for radiation pattern measurement.

summarizes the *HPBW* of the array that shows that, for all cases, the measured *HPBW* are smaller than the simulation result.

5.3.7.4 *Gain Measurement* The measurement of the array gain is a combination of measurement, simulation, and calculation process. The standard horn antenna is used as transmit antenna and AUT receives the signal, while the Agilent PNA measures the S_{21} of the link. The gain of the array relates to other parameters by

$$S_{21} = P_{Tx} - P_{Rx} = G_{Tx} + G_{AUT} - L_{cable} - L_{FS} - L_{Feedline} \text{ all in dB} \quad (5.9)$$

where G_{Tx} is the gain of horn antenna, which is 10.6 dB$_i$. L_{cable} is the cable loss and can be measured directly with the Agilent PNA E8361A using full two-port error correction calibration. In measuring L_{cable}, the loss of connectors and adaptors are also considered. Free space loss or path loss, LFS, is calculated using Friis transmission formula [41]. The feedline loss should be considered as the CPW feed extension is 4λ long and contributes significantly to the gain result. This loss has been separately simulated in CST and shown in Figure 5.28. The feedline is primarily designed at 60.5 GHz, which is the center frequency of 57–64 GHz. Therefore, the feedline shows the minimum insertion loss at 60.5 GHz and more loss in other frequencies of the band. The 1.85 mm Southwest Microwave connector has a VSWR of 1.25 [44] means that the impedance mismatch of the connector should not have significant effect on the gain measurement. Therefore, the connector's performance is not excluded from the gain measurement. By considering Equation (5.9) and by

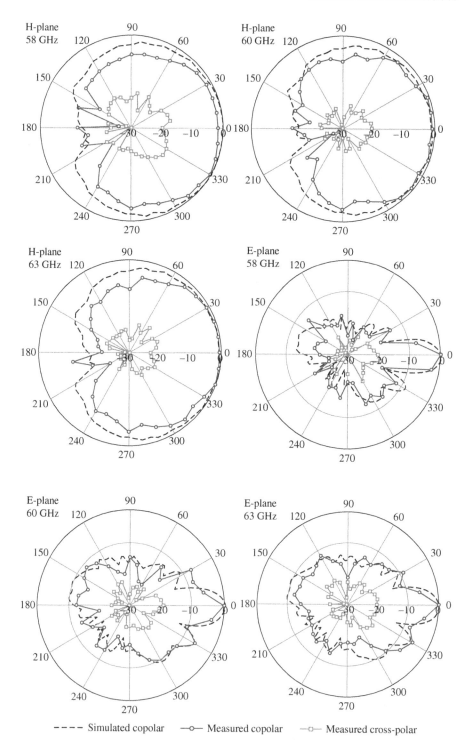

Figure 5.26   Measured and simulated E-plane radiation pattern, co- and cross-polar radiation.

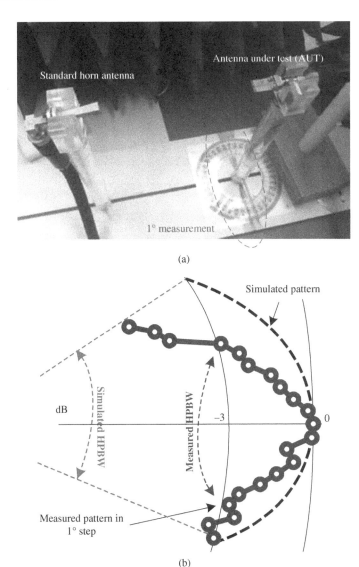

Figure 5.27   (a) Measurement setup for 1° accuracy and (b) measured and simulated *HPBW* of E-plane.

combining the measured data with the LFS and simulated $L_{feedline}$, the final value of the gain versus frequency is shown in Figure 5.28. As can be seen from the figure, the feedline loss and antenna gain are congruent to each other over the frequency band. There is about 1.5 dB gain drop from 7 dB$_i$ due to 1.5 dB insertion loss variation of the feedline. The absolute gain of the array antenna varies from 5.5 to 7 dB$_i$ over the frequency band of operation.

TABLE 5.6   Simulated and Measured *HPBW* in E-Plane

Frequency (GHz)	E-Plane (°)		H-Plane (°)	
	Simulated	Measured	Simulated	Measured
58	21	15	182	135
60	18	15	185	135
63	17	13	190	150

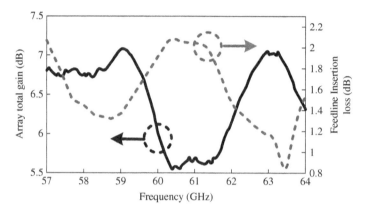

Figure 5.28   Measured gain and simulated feedline loss of array.

5.4   CONCLUSIONS

This chapter discussed about some key technical specifications of the proposed spatial-based system for the chipless RFID system. First it suggested the
millimeter-band 60 GHz as the appropriate band for precise scanning of the tag.
Then the general situation of the unlicensed 60 GHz band has been reviewed from a
technical and regulatory perspective. The low regulatory restrictions on the EIRP of
this band have been highlighted that makes the band very suitable for the proposed
technique. Moreover, it was shown that the minimum amount of metal is required
for the printed tag at this band, which lowers the tag price below that for the barcode.

In the second part of the chapter, the technical and operational necessities of the
reader antennas for the proposed technique were reviewed. Based on the proposed
application, the technical specifications of the reader antenna were established first.
A survey on the available open source products was conducted and their technical
specifications were reviewed. The process of the antenna design was then introduced.
An array of DSPD antennas was proposed to satisfy the demanded technical specifications. A DSPD array antenna was designed and thoroughly investigated at 60 GHz.

First, the Taconic TLX-8 was characterized using an SIW resonant technique.
The measured constitutive parameters at 60 GHz significantly differed from that of

published data at 10 GHz. The new values were used for designing a four-element DSPD array antenna. In the design process, the following advanced design techniques were used: (i) a broadband matching technique using dipole arms chamfering and abrupt ground plane for the dipole element, (ii) element array analysis and mitigating mutual coupling between elements, (iii) beam-forming network with matching stub at the T-junction, (iv) microstrip-to-CPW transition and SIW-based CPW extensions, and finally (v) a V-type connector assembly. The array of four DSPD elements covered the whole frequency range of 57–64 GHz yielding 12% bandwidth. While the array dimension was very small and the standard printed circuit board technology was used for the antenna fabrication, the measurement results were well matched with the simulation predictions. The measured *CPL* was below -20 dB as required for proper operation of the spatial-based system. The measured gain of the array was 5.5–7 dB$_i$ over the entire frequency range of operation. The size of the array was 22×14 mm^2. The compact and simple structure of the array also made it suitable for many applications in the millimeter-wave region for low-range wireless communication systems.

REFERENCES

1. ITU-R, "Rec.p 676-9, Attenuation by Atmospheric Gases," International Telecommunication Union, Geneva, 2012.

2. C. Koh, "The Benefits of 60 GHz Unlicensed Wireless Communications," YDI Company, USA.

3. ITU, *Radio Regulation*, ITU, Geneva, 2012.

4. S.K. Yong, P. Xia, and A.V. Garcia, *60 GHz Technology for Gbps WLAN and WPAN: From Theory to Practice*. John Wiley & Sons UK, 2010.

5. N. Guo, R.C. Qiu, Sh.S. Mo, and K. Takahashi, "60-GHz Millimeter-Wave Radio: Principle, Technology, and New Results," *EURASIP Journal on Wireless Communications and Networking,* vol. 2007, pp. 1–9, 2007.

6. ACMA, "60 GHz Band, Millimetre Wave Technology," 2004.

7. S. Yong and C. Chong, "An Overview of Multi Giga bit Wireless through Millimeter Wave Technology: Potentials and Technical Challenges," *EURASIP Journal on Wireless Communications and Networking,* vol. 2007, p. 10, 2007.

8. S. Yong, P. Xia, and A.V. Garcia, *60 GHz Technology for Gbps WLAN and WPAN*, John & Wiley, 2010.

9. W. Buchar, N. Karmakar, and M.Zomorrodi, "MIMO-Based Technique for Chipless RFID EM-Imaging at 60 GHz," Grant proposal Xerox, USA Sep 2014.

10. HURZ Stamping Technology, (Feb 2015). Available: http://www.kurz.de.

11. ACMA, "LIPD Regulation," 2011.

12. ACMA, "Radiocommunications (Low Interference Potential Devices) Class Licence 2000," ACMA, Canberra 2011.

13. Comlab Inc., Nokia. (14 Oct 2014). *Radio Link Systems Nokia MetroHopper 58 GHz*. Available: www.comlab.hut.fi/studies.

14. Proxim Wireless Inc., Oct 2014). *Terabeam Product*. Available: http://www.proxim.com/.

15. A. Nesic, S. Jovanovic, and V. Brankovic, "Design of Printed Dipoles Near The Third Resonance," in *Antennas and Propagation Society International Symposium*, Atlanta, GA, USA, 1998, pp. 928–931.

16. R.B. Waterhouse, D. Novak, A. Nirmalathas, and C. Lim, "Broadband Printed Sectorized Coverage Antennas for Millimeter-Wave Wireless Applications," *IEEE Transactions on Antennas and Propagation,* vol. 50, pp. 12–16, 2002.

17. W. Menzel, "A 40 GHz Printed Array Antenna," in *Digest*, pp. 225–226, 1980.

18. C. Peixeiro, P. Dufrane, and Y. Gullerme, "Microstrip Patch Antennas for a Mobile Communications System at 60 GHz," in *Antennas and Propagation Society International Symposium, AP-S, Digest*, Baltimore, MD, USA, 1996.

19. T.G. Ma and S.K. Jeng, "A Printed Dipole Antenna with Tapered Slot Feed for Ultrawide Band Applications," *IEEE Transactions on Antennas and Propagation,* vol. 53, pp. 3833–3839, 2005.

20. M. Scott, "A Printed Dipole for Wide-Scanning Array Application," in *IEEE 11th International Conference on Antennas and Propagation*, 2001, pp. 37–40.

21. Y.H. Suh and K. Chang, "A New Millimeter-Wave Printed Dipole Phased Array Antenna Using Microstrip-Fed Coplanar Stripline Tee Junctions," *IEEE Transactions on Antennas and Propagation,* vol. 52, pp. 2019–2027, 2004.

22. V. Brankovic and A. Nesic, "Wide Band Printed Phase Array Antenna for Microwave and mm-Wave Applications," US Patent, March 2000.

23. R.A. Alhalabi and G.M. Rebeiz, "High-Efficiency Angled-Dipole Antennas for Millimeter-Wave Phased Array Applications," *IEEE Transactions on Antennas and Propagation,* vol. 56, p. 7, 2008.

24. Y.C. Chiou, R.A. Alhalabi, and G.M. Rebeiz, "High-Efficiency 60 GHz Dipole-Box Antennas," in *Antennas and Propagation Society International Symposium (APSURSI)*, 2010.

25. P. Kalansuriya, N.C. Karmakar, and E. Viterbo, "On the Detection of Frequency-Spectra-Based Chipless RFID Using UWB Impulsed Interrogation," *IEEE Transactions on Microwave Theory and Techniques,* vol. 60, pp. 4187–4197, 2012.

26. Taconic, "Taconic Laminate Material Guide," 2014.

27. R. Anita and M.V.C. Kumar, "Analysis of Triangular Microstrip Patch Antenna for Different Antenna," in *JREAT International Journal of Research in Engineering & Advanced Technology*, 2013.

28. H.R. Fettermana, T.C.L.G. Sollnera, P.T. Parrisha, D. Parkera, H. Mathewsa, and P.E. Tannenwalda, "Printed Dipole Millimeter Antenna for Imaging Array Applications," *Electromagnetics,* vol. 3, pp. 209–215, 1983.

29. D.M. Pozar, "Considerations for Millimeter Wave Printed Antennas," *IEEE Transactions on Antennas and Propagation,* vol. 31, No. 5, p. 8, 1983.

30. P. Katehi and N. Alexopoulos, "On the Effect of Substrate Thickness and Permittivity on Printed Circuit Dipole Properties" *IEEE Transactions on Antennas and Propagation,* vol. 31, p. 6, 1983.

31. W. Hong and K. Wu, "94 GHz Substrate Integrated Monopulse Antenna Array," *IEEE Transactions on Antennas and Propagation,* vol. 60, pp. 121–129, 2011.

32. J. Sheen, "Comparisons of Microwave Dielectric Property Measurements by Transmission/Reflection Techniques and Resonance Techniques," *Measurement Science and Technology,* vol. 20, pp. 1–12, 2009.

33. D.E. Zelenchuk, V. Fusco, G. Goussetis, A. Mendez, and D. Linton, "Millimeter-Wave Printed Circuit Board Characterization Using Substrate Integrated Waveguide Resonators," *IEEE Transactions on Microwave Theory and Technique*, vol. 60, 3300–3308, 2012.

34. M. Bozzi and L. Perregrini, "Modeling of Conductor, Dielectric, and Radiation Losses in Substrate Integrated Waveguide by the Boundary Integral-Resonant Mode Expansion Method," *IEEE Transactions on Microwave Theory and Techniques,* vol. 56, pp. 3153–3161, 2008.

35. W.L. Stutzman and G.A. Thiele, *Antenna Theory and Design*, John Wiley & Sons, 1981.

36. G.Y. Chen and J.Sh. Sun, "A Printed Dipole Antenna with Microstrip Tapered Balun," *Microwave and Optical Technology Letters,* vol. 40, pp. 344–346, 2004.

37. R.J.P. Douvilie and D.S. James, "Experimental Study of Symmetric Microstrip Bends and Their Compensation," *IEEE Transactions on Microwave Theory and Technique*, vol. 26, p. 175, 1978.

38. S. Preradovic, "Chipless RFID System for Barcode Replacement," PhD, ECSE, Monash University, 2009.

39. S.M. Roy, "Development of a Frequency Encoded Chipless RFID Tag," PhD, ECSE, Monash University, 2008.

40. R.L. Haupt, *Antenna Arrays, A Computational Approach*. Pennsylvania State University, State College, John Wiley&Sons, Inc., Publication, Pennsylvania, 2010.

41. C.A. Balanis, *Antenna Theory Analysis and Design*, Third ed.. John Wiley & Sons, Hoboken, New Jersey, , 2005.

42. D. M. Pozar, *Microwave Engineering*, Third ed., John Wiley & Sons, 2005.

43. J. Coonrod, "Comparing Microstrip and CPW Performance," *Microwave Journal,* vol. 55, pp. 74–82, 2012.

44. Southwest Microwave, Inc., S. Microwave. "1.85 mm Connectors". Available: www .southwestmicrowave.com/.

45. (2013). *SouthWest Microwave*. Available: www.southwestmicrowave.com.

46. Ainfoinc Company, (14 March 2012). *Ainfoinc Company*. Available: http://www.ainfoinc .com.

6

SAR-BASED SIGNAL PROCESSING

6.1 INTRODUCTION

It was already discussed that it is possible to provide a fine spatial diversity for data encoding purpose in a chipless radiofrequency identification (RFID) system by utilizing the millimeter-wave bands of 60 GHz and above. However, considering the reading range, tag size, and expected data encoding capacity, it is not technically possible to use one single antenna at the reader and expect the millimeter resolution on the tag surface. Instead, a synthetic aperture length of 15–30 cm can be constructed by a small moving antenna. The signal processing based on the synthetic aperture radar (SAR) technique then provides a spatial diversity of millimeter order at the tag surface.

This chapter reviews the SAR-based signal processing for the proposed chipless application. Given that the SAR technique is a mature field of research with very wide applications, the very basics of the SAR will be reviewed here and readers can refer to the appropriate books in this field [1]. The spotlight mode of operation will be applied for the proposed application for tag data decoding purpose.

First a sample tag with only 4 data bits capacity is considered for the proof of concept. The effect of synthetic aperture size on the achievable spatial resolution is discussed as well. In the final section of this chapter, the ultimate encoding capacity of the proposed technique is considered based on the maximum allowable tag surface. A tag that is 5.5 times smaller than a credit card encodes 17 bits of data, while it is fully printed by a SATO printer. This data encoding capacity based on a fully printed structure would be seen as a revolutionary step forward achieving EPC global standard from millimeter-wave chipless RFID systems.

Advanced Chipless RFID: MIMO-Based Imaging at 60 GHz−ML Detection, First Edition.
Nemai Chandra Karmakar, Mohammad Zomorrodi, and Chamath Divarathne.
© 2016 John Wiley & Sons, Inc. Published 2016 by John Wiley & Sons, Inc.

6.2 SAR MODES OF OPERATION

SAR-based systems may work in different modes of operation that depends on the system flexibilities and the application requirements. Three main modes of operation are (i) spotlight, (ii) strip map, and (iii) scan mode [2]. In the spotlight mode, the radar system steers its antenna toward a certain area where the target is located. The spotlight mode is normally used when the target area is small and a fine image resolution of the target is required. In the strip map mode, the radar antenna is fixed with respect to the system and the moving route. These two modes of operation are shown in Figure 6.1. The strip map mode is useful for providing a coarse image resolution of a large area of interest, earth imaging, for example. If these two modes are combined, then the SAR system works on the scan mode. The azimuth resolution provided by any mode of the SAR system directly depends on the length of the synthetic aperture size. In the strip map mode, the synthetic aperture length is fixed as the radar antenna has a fixed angle with the airplane. On the other hand, the spotlight mode does not experience any limitations on its synthetic aperture length as the system has the ability to steer the beam of the physical antenna and illuminate the scene for a longer period.

For the proposed millimeter-wave chipless RFID tag, the maximum area the tag occupies is less than a credit card size (85×55 mm^2). On the other hand, the maximum azimuth resolution is desirable as it defines the tag's data encoding capacity. Therefore, the spotlight mode of SAR seems to be an appropriate mode of operation for the RFID tag imaging. Through the curvature nature of the reader route around the tag, it is possible to focus the radiation pattern of the reader antenna on the tag surface and illuminate the tag through different view angles. Moreover, flexibility on the length of the synthetic aperture size at the spotlight mode results in a finer azimuth resolution at the expense of a larger synthetic aperture length.

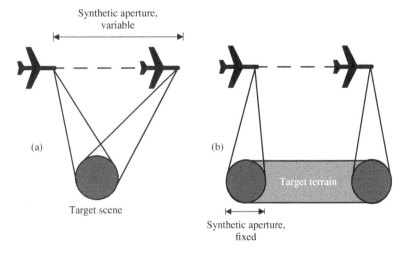

Figure 6.1   SAR modes of operation: (a) spotlight mode and (b) strip map mode.

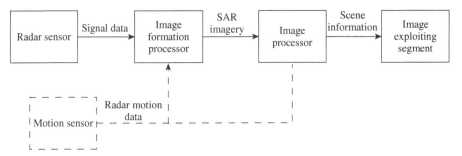

Figure 6.2   General block diagram of typical SAR-based system [2].

6.3   SAR BLOCK DIAGRAM

This section reviews the fundamentals of the SAR system and introduces basic expression describing the spotlight mode as the more appropriate mode of operation for the proposed RFID application. The block diagram of a typical SAR-based system is shown in Figure 6.2. The radar antenna provides real-time data of the scene. In many cases, the information about the movement of the radar system is collected during the operation of the system. These two sets of data provide the scene data as well as the complementary data of the radar motion. The *image formation processor* uses these two data sets to provide the SAR imagery to the *image processor* and subsequently to the *image exploitation segment* [2]. For many applications, including earth imaging, the movement of the radar system is almost a fixed path; however, due to unexpected factors such as wind and the terrain of the earth, small variations of the radar route often happen. Therefore, the motion sensor is an unavoidable part of the system. For the RFID application, however, the movement is completely predictable as the radar moves on a rail. Therefore, this section can be ignored, while the information of the route would be easily considered mathematically in the subsequent modules. A detailed description of the modules in Figure 6.2 is described in Ref. [2].

6.4   SAR-BASED SIGNAL PROCESSING

In this section, the image of the tag is constructed through SAR technique. The tag is comprised of five meander lines as EM polarizers in each column for data encoding purposes. Only fine azimuth resolution is targeted through the SAR technique and no attempt is made on the range resolution because of the limited available frequency spectrum. The signal processing fundamentals are based on the SAR technique for azimuth resolution and the range migration algorithm (RMA) for the range resolution as described in Ref. [2]. Moreover, Charvat [3] simplified the proposed approaches in Ref. [2] for low-range radar systems. The current work follows the Carrara *et al.* [2] and Charvat [3] approaches with appropriate modifications for the proposed application of the chipless RFID systems in millimeter-wave band.

6.4.1   System Structure for Data Collection

The proposed structure of the tag is already shown in Figure 3.11. The described antenna in the previous chapter, which covers the frequency band 57–64 GHz, moves on a straight line around the tag. The straight line that is related to the strip map mode is assumed for simplicity of the signal processing. However, for the final stage, the curved trajectory may be considered that provides better azimuth resolution as it uses the benefits of the spotlight mode of SAR operation. The read range may vary from 10 to 50 cm. However, all the results shown in this chapter are based on 10 cm reading distance due to the transmit power limitation of the PNA at the 60 GHz band. This read range is in line with all measurements carried out so far and shown in the previous chapters. The reading step over the synthetic aperture is 2 mm that is equivalent to almost 0.4λ. This means that the reader illuminates the tag surface in every 2 mm and collects the backscattered signal and then moves to the next position. The synthetic aperture size may vary from 15 to 30 cm. However, for 10 cm reading distance, the length of 20 to 25 cm is normally enough for the synthetic aperture size to achieve 5 mm spatial resolution.

Considering the above-mentioned technical parameters, the reading process of the tag requires tremendous number of separate transmissions and receptions of the signal for data collection purposes; 125 steps for a 25 cm aperture size, for example:

$$\text{Number of readings} = \frac{\text{aperture size}}{\text{reading step}} = \frac{250\,\text{mm}}{2\,\text{mm}} = 125 \qquad (6.1)$$

This seems to be very slow for data measurement and may be counted as a big disadvantage of the proposed technique. This issue is addressed in the following chapter.

For the range resolution, a normal chirp signal with FM modulation is assumed. The chirp rate of 400 GHz/s, a nominal value for a chirp signal generator, is considered. At the millimeter-wave band, the start frequency of 58.2 GHz and the stop frequency of 62.2 GHz are assumed. The pulse duration is 10 ms; hence, the occupied frequency band of the chirp is 4 GHz. Considering the relation between the range resolution and the occupied bandwidth, one may expect a very coarse range resolution in multicentimeter order that would be much rough resolution compared with the azimuth resolution. This means that the effects of all five meander-line polarizers on each strip of the tag surface are combined and cannot be separated by the reader.

6.4.2   Signal Processing Steps

As mentioned earlier, the RMA algorithm is followed for the SAR-based coding. The captured data based on various positions of the reader antenna, known as the raw signal, is processed by the RMA SAR-based algorithm. The RMA is normally divided into the following separate steps:

- Data collection
- Azimuth Fourier transform

- Matched filter
- Stolt interpolation
- Inverse Fourier transform
- Reflectivity image.

During the movement of the reader around the tag, a 2D data matrix is produced, which contains the range profile data. Hence, the collected data has the following formation:

$$S(x_i, w(t)) \tag{6.2}$$

where x_i contains the information of the reader position and the $w(t)$ is the instantaneous of the chirp signal and defined by Equation (6.3) while the c_r is the chirp rate, f_c is the center frequency of the signal, and BW is the signal bandwidth:

$$w(t) = 2\pi \left(c_r t + f_c - \frac{BW}{2} \right) \tag{6.3}$$

The matrix format of Equation (6.2) is considered as the raw signal by the codes for SAR-based signal processing. In the RMA approach, the azimuth Fourier transformation is applied to this raw signal. This is the first difference between the RMA and other conventional approaches in SAR techniques, for example, polar formation algorithm. The Fourier transformation changes the time-domain signal, $S(x_i,w(t))$, to its equivalent wave number domain, $S(k_x,k_r)$.

The next phase of the RMA is to apply the 2D phase compensation to the azimuth-transformed signal [2]. This operation perfectly corrects the range curvature of all scatterers at the same range as the scene center. The matched filter is defined through the following expression in which the R_s shows the distance between to the target center [2,3]:

$$S_{mf}(k_x, k_r) = e^{jR_s}\sqrt{k_r^2 - k_x^2} \tag{6.4}$$

Following the matched filter step, Stolt interpolation is applied in the RMA algorithm. It simultaneously compensates the range curvature of all scatterers by an appropriate wrapping of the SAR signal data. This means mapping of the 2D SAR data from the wave number domain (k_r) to the spatial wave number domain k_y through the following expression [2]:

$$k_y = \sqrt{k_r^2 - k_x^2} \tag{6.5}$$

A 1D interpolation must be conducted across all the azimuth wave number k_r to map them onto k_y thus resulting in Stolt-interpolated matrix $S_{st}(k_x,k_y)$ [3].

At this point, the processed signals from all scatterers are 2D linear phase gratings. A 2D inverse Fourier transform is computed to fully compress the scatterers in range and azimuth [2]. The above-mentioned process is illustrated in Figure 6.3.

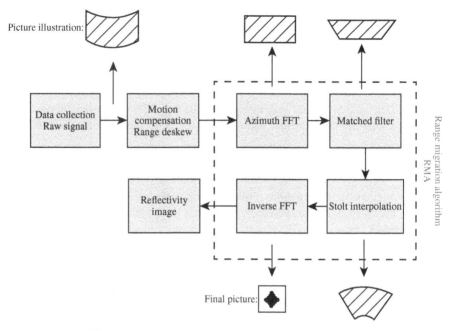

Figure 6.3   Image formation processing using RMA [2].

6.5   TAG IMAGING RESULTS

The measurement setup is similar to what explained in the previous chapters. Two sets of the designed array of digital signal processing devices (DSPDs) antennas are used as Tx and Rx antennas. The antennas are connected to the PNA while oriented orthogonally. The printed tag is located at 10 cm distance to the antennas as at the frequency band 57–64 GHz, the maximum leveled power of PNA is −5 dBm. Considering the maximum gain of array antenna, 7 dB$_i$, the EIRP of the system is only 2 dBm, while the maximum EIRP of 40 dBm is allowed at this band [4,5]. Therefore, the reading distance may increase to 50 cm in commercial phase of the proposed EM-image chipless RFID system if a higher power signal source is available at 60 GHz.

6.5.1   Proof of Concept

Showing the applicability of the proposed technique for data encoding of a printed EM-polarizer-based tag, two tags with different encoded data are considered. The tags only include 4 bits of data, "1001" and "1011." They are 2 cm long while their width is fixed to almost 1 cm comprising five meander lines in each column. A sample tag with 4 bits of data is shown in Figure 6.4.

These two tags are interrogated through the reader at different angles while the reader moves on a straight line of 22 cm. It is important to note that the length of the

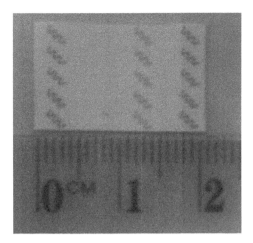

Figure 6.4 Photograph of 4-bit printed tag.

synthetic aperture is 20 cm; however, to scan the total tag's length, it is required to add the length of the tag to the initial synthetic aperture length. The received signals before SAR-based signal processing, the raw signal, are shown in Figure 6.5. The received power depends on the exact frequency of the chirp signal, but the values shown in Figure 6.5 are the normalized power for two tags "1001" and "1011," respectively. As it is clear, these raw signals convey no information without proper signal processing. However, comparing the two figures in Figure 6.5 reveals that only changing 1 bit of the tag's data completely changes the pattern of the received signal.

Based on the images in Figure 6.7, the azimuth spatial resolution is 5 mm, which is clearly shown in the processed images. This is the real azimuth resolutions so far obtained. Therefore, the total area that each column occupies is 0.5 cm^2 considering the 1 cm length of each column. As each column may represent "1," "0," or one binary bit, it can be concluded that the total data encoding capacity is 2 bits/cm^2 or equivalently four different states on each square centimeter. This is the data encoding capacity that has been achieved so far. However, the authors believe that with more advanced signal processing techniques, the achievable data encoding capacity may increase. The total data capacity of the tag directly depends on the tag's length. Each 0.5 cm of the tag length encodes 1 bit of data and for increasing the encoded data, the length of the tag needs to increase.

It is important to mention that the vertical axis that relates to the range resolution is not the real range resolution of the processed EM image. It is actually assumed in the signal processing that all of the five polarizers are located at the center point of the strip. However, the actual range resolution is much higher than the values shown in Figure 6.7.

The values on the color bar of Figure 6.7, which is congruent to the signal level in Figure 6.6, show the difference between the signal level, in the cross-polar direction, from a strip with five EM polarizers, and the reflection from other parts of the tag without any polarizer. These values are matched with what was experienced in the

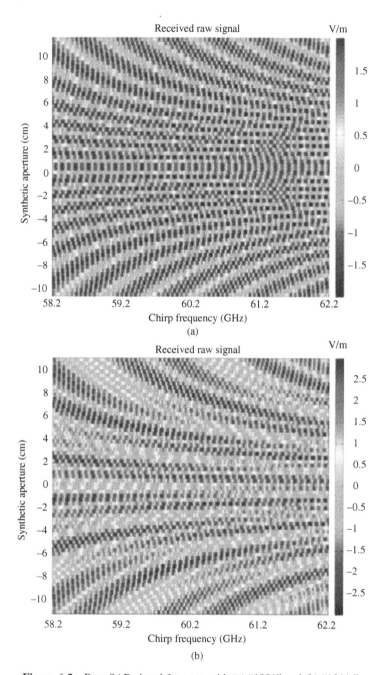

Figure 6.5   Raw SAR signal for a tag with (a) "1001" and (b) "1011."

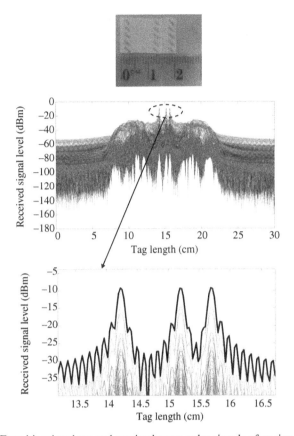

Figure 6.6   Four-bit printed tag and received cross-polar signals after signal processing.

measurement process in the previous chapters. One may recall that five printed meander lines created 15 dB higher cross-polar component compared with the no tag situation.

6.5.2   Image Resolution Versus Aperture Length

The decoded tags in the previous section were read at 10 cm distance with the synthetic aperture size of 22 cm that yielded 5 mm image resolution. However, Figure 2.9, which relates the aperture size to the spatial resolution, suggests shorter lengths for the 5 mm image resolution at the reading distance of 10 cm. To verify that the 20 cm synthetic aperture length is the minimum length, which can adequately decode the tag data, two lengths of 12 and 17 cm are assumed for the synthetic aperture. The images of the tag with "1011" data are shown in Figure 6.8 for the aperture lengths 12 and 17 cm, respectively.

As is clear from Figure 6.8, the provided image by 12 cm aperture size does not decode the tag's data as "1011." The effect of each column of the polarizers has spread

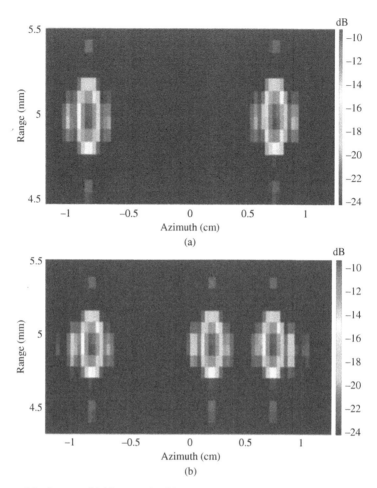

Figure 6.7   Image of 4-bit tags with 22 cm aperture size, (a) "1001" and (b) "1011."

over the other segments of the image that is not matched with the initial size of the tag. The image in Figure 6.8 has almost 5 cm length, while the initial tag is only 2 cm long. Hence, the image resolution of 5 mm is no longer existed for the case of 12 cm aperture size.

Increasing the aperture size to 17 cm enhances the final image resolution on which the effects of three columns of polarizers are detectable as shown in Figure 6.8. However, the final azimuth resolution is less than what is already shown in Figure 6.7. Again, the image of the tag spreads to 3 cm, while the initial tag is only 2 cm. Moreover, the vertical position of the column of polarizers in Figure 6.7 is very clear and matches the actual position of the meander lines on the tag surface. While in Figure 6.8, the image shows inclined rows of polarizers, which makes it difficult to separate the effect of the adjacent columns of polarizers. Therefore, for the aperture size of 17 cm, the data decoding is very difficult if not impractical.

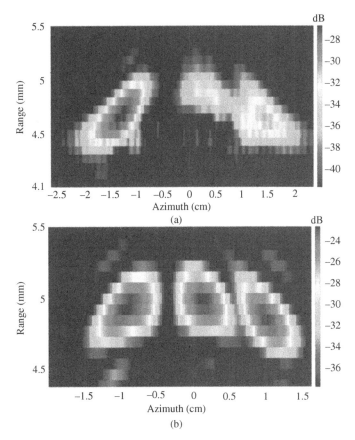

Figure 6.8 EM image of the tag "1011" with (a)12 cm and (b) 17 cm aperture size.

Three reasons can be suggested for the difference between the predictable image resolution and the actual image resolution after signal processing. First, the graphs in Figure 2.9 are based on reflector antenna formula [6] while in the measurement process the reader travels on a straight line. Moreover, the predicated synthetic aperture is related to a physical antenna, while in the signal processing a finite number of sample points exist to simulate an array antenna structure. The differences between these two ideas are qualitatively shown in Figure 2.8. Finally, the utilized signal processing algorithm is the standard SAR signal processing approach. There are advanced SAR-based signal processing techniques that enhance the final image resolution. It is also important to mention that the final aperture length for imaging the whole tag is equal to the fundamental synthetic aperture length for a certain resolution plus the length of tag. This issue is figuratively shown in Figure 6.9 and also some sample aperture lengths are mentioned in Table 6.1. As one may notice, the initial fundamental aperture length of less than 20 cm is not long enough for targeted 5 mm image resolution.

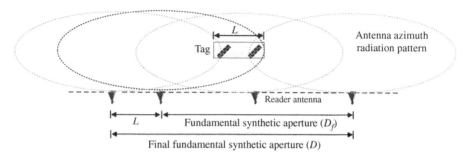

Figure 6.9   Final and fundamental aperture length.

TABLE 6.1   Synthetic Aperture Size and Image Resolution

Minimum Final Aperture Length (cm)	Fundamental Aperture Length (D_f) (cm)	Image Resolution (mm)	Tag Length (cm)	Tag Data Capacity	Reading Process
12	10	>5	2	4	Not successful
17	15	>5	2	4	Not successful
22	20	5	2	4	Successful
24	20	5	4	8	Successful
25.5	20	5	5.5	11	Successful
28.5	20	5	8.5	17	Successful

6.5.3   High-Data Content Tags

In the previous section, the concept of EM imaging for data encoding purposes was presented and approved. It was shown that a 4-bit tag can be read and decoded at 10 cm reading distance through its resultant image after SAR signal processing. It is therefore of utmost benefit to explore the final content capacity of the proposed system.

An 8-bit tag is first tested by the proposed technique. The tag at 10 cm distance is interrogated by the reader antennas. The tag structure, received raw signal, received signal after signal processing, and finally the tag's EM image are shown in Figure 6.10. The received signal level after signal processing is highlighted for better matching with the tag's image. As suggested, the tag length is 4 cm and its width almost 1 cm. The final image of the tag clearly shows the position of each column of five polarizers on the tag surface. There is no overlay between adjacent columns. The SAR aperture size is $(20 + 4)$ cm. A total of 20 cm is required for 5 mm azimuth resolution and 4 cm will be added due to the total length of tag.

Figure 6.11 shows two tags with 11 and 17 data bits, respectively, along with their final EM image. The raw signals and the actual signal levels after signal processing have not been shown for these two cases. Again the aperture size for each case is approximately the initial aperture length 20 cm plus the length of tag. As is clear

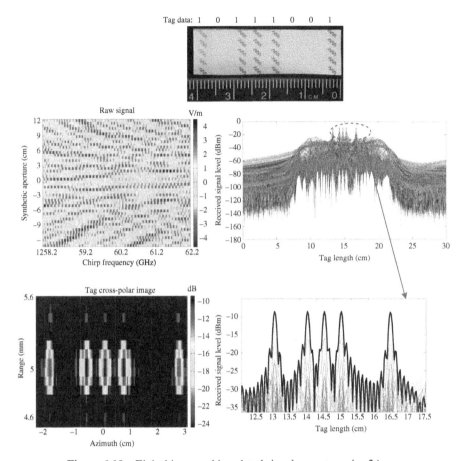

Figure 6.10 Eight-bit tag and its related signals, aperture size 24 cm.

from the related EM images, the processed image successfully reveals the presence or absence of polarizers on each column of the tag surface, hence data decoding is performed successfully. These results confirm the validity of the proposed theory at its maximum performance capability. As discussed earlier, the acquired cross-range resolution is again 5 mm and the width of the tag is around 10 mm with five polarizers on each column.

For a chipless tag of $8.5 \times 1 \text{ cm}^2$, 17 bits can be encoded as shown in Figure 6.11. This tag is 5.5 times smaller than a credit card ($8.5 \times 5.5 \text{ cm}^2$). Hence, it can be concluded that in an area equivalent to a credit card, 93-bit encoding capacity is feasible. This encoding capacity for a chipless RFID system is very attractive. The acquired data encoding capacity is based on a fully printed tag structure. The printed tag is fabricated on a paper substrate with the minimum conductive ink usage while it suffers from many structural anomalies. The robustness of the proposed tag structure with respect to multipath and clutter interferences has been already shown through

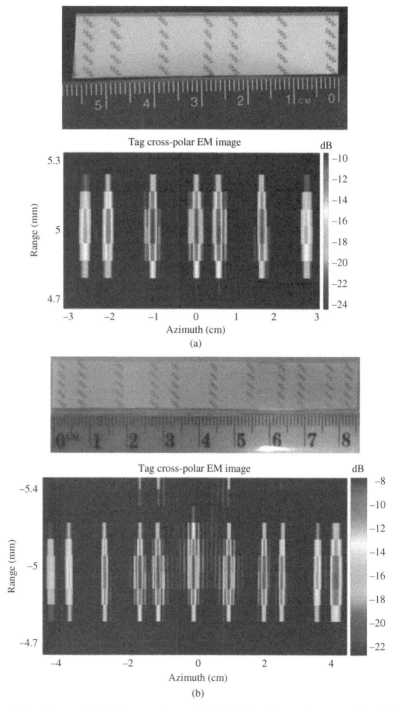

Figure 6.11   Cross-polar EM image of tag with (a) 11 bits of data and aperture size 25.5 cm and (b) 17 bits of data and aperture size 29 cm.

measurements. The chipless tag is also applicable for tagging of metallic and high reflective items. The tag is also concealable inside many items suggesting the opportunity for secure identification. Therefore, it can be concluded that a very successful and innovative approach has been suggested for the chipless RFID systems.

6.6 SYSTEM DOWNSIDES

The proposed spatial-based system suggests an interesting and effective approach for the chipless tags with many salient attributes. However, there are two limitations associated with the proposed EM-imaging technique that may deter its full potentials in mass deployment of the chipless RFID systems. The main issue associated with the proposed EM-imaging technique is the requirement for relative movement of the reader and tag has been as already addressed. The second limitation is the sensitivity of the system to the tag orientation.

6.6.1 Moving Reader

EM imaging of the tag's surface through SAR requires relative movement of the reader antenna with respect to the tag. As the tag is normally fixed or its moving trajectory is not predictable, a fixed tag and a moving reader are assumed to be the most general scenario in the proposed system. As discussed before, illumination of the tag and gathering its backscattered signal from different view angles normally require large number of individual transmit and receive scans, in order 100–150 times. This severely affects the system's reading speed that may be seen as the main drawback of the proposed approach. The movement of the reader around the tag for the SAR processing will not be based on the operator's hand movement. Obviously, the speed, aperture length, and other factors cannot be maintained accurately for SAR signal processing through manual movement. Instead, the moving reader antenna will be encompassed in a jig and the antenna moves over a rail of maximum 30 cm length while controlled by a stepper motor as shown in Figure 6.12. Therefore, the moving reader antenna collects the backscattered signals while continuously moving around the tag, similar to the earth imaging approach. In this scenario, a Doppler frequency shift due to continuous movement of the reader antenna is expected. Although this may alleviate the long process of the measurement, mechanical movement of the antenna is not the efficient way and would not be accepted in the modern world.

One may suggest using of array or phase array antennas. Considering the required pattern precision for the SAR signal processing and the technical parameters of the proposed system including its frequency band of operation and operational characteristics, the array antenna solution would be very complex and expensive and hence not practical. Instead, the new technique of multiple input multiple output (MIMO) can be combined with the SAR approach and provide a practical solution with minimum hardware complexity. This technique will be discussed in detail in next chapter.

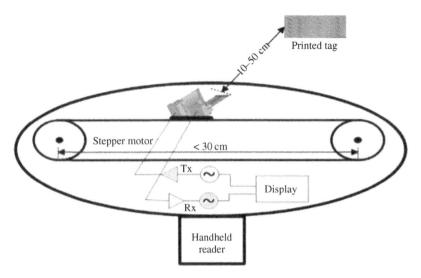

Figure 6.12 Handheld reader with mechanically moving antennas.

6.6.2 Tag Orientation Sensitivity

The proposed approach of the spatial-based system is initiated on the cross-polar component basis. As disclosed earlier, the strip line and meander line act as effective polarizers if the incident E-field maintains 45° angle with the polarizer's main axis. Otherwise, the performance of these EM polarizers would be different than what is normally expected. In this regard, all the measurements, simulation, and signal processing results shown in the previous chapters are based on 45° orientation between the EM polarizers and the incident E-field as shown in Figure 6.13(a). To evaluate the effect of tag misorientation on the system performance, the polarizer on the tag surface is rotated at an arbitrary angle θ as shown in Figure 6.13(b). Various oriented tags are simulated and the results are shown in Figure 6.14. It is clear from the figure that the tag misorientation affects the received signal level. This is due to the reduction of the resonance factor on the cross-polar components. It is important to mention that the value in Figure 6.14 shows the received E-field level for one single EM polarizer. The related power reduction hence is double what is shown in Figure 6.14. As shown before, all the measurements were based on the 12–15 dB difference power level between backscattered signals from the polarizers. Therefore, for tag orientation angles more than 20°, the received signal level drops to more than 5 dB and hence the system ability to detect the column of EM polarizers will be severely affected as shown in Figure 6.15.

The suggested solution for providing more robust system to the tag orientation is the usage of circular polarization rather than the linear polarized interrogation signal. The entire proposed theory was based on the linear polarization scheme of the Tx and Rx antennas at the reader side. The EM polarizer on the tag surface rotates the linear polarization to its orthogonal direction. However, if the incident signal has

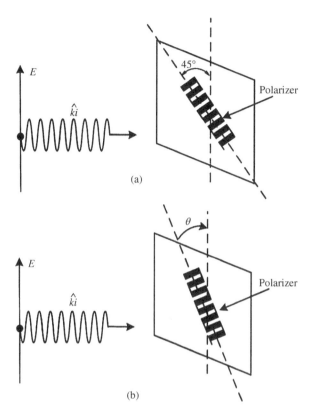

(a)

(b)

Figure 6.13 Tag orientation: (a) correct angle of 45° and (b) misoriented tag angle of $\theta°$.

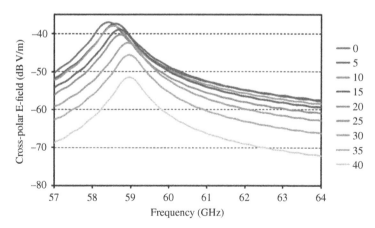

Figure 6.14 Effect of tag orientation on received signal level.

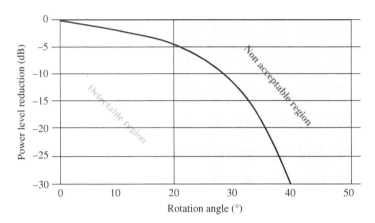

Figure 6.15   System degradation due to tag misorientation effect.

circular polarization, the EM polarizer rotates it to the orthogonal direction. This means that a right-hand circular polarization (RHCP) is converted into the left-hand circular polarization (LHCP) and vice versa. This provides insensitive system with respect to the tag rotation. However, it should be mentioned that the range and azimuth resolutions of the final tag's EM image will be also considered with this new system arrangement. This means that the definition of the range and azimuth resolutions will be different and hence a new configuration is required. Obviously, the designed array of DSPD would not be suitable anymore for the new circularly polarized (CP) system and hence new CP antennas are required.

6.7   CONCLUSIONS

The synthetic aperture radar theory has been briefly reviewed in this chapter. Three modes of operation of the SAR theory along with their attributes and limitations were discussed. Although the strip map mode is applicable for the proposed application, the spotlight mode is able to provide better azimuth resolution due to the increased length of the synthetic aperture size.

Some sample tags with different data encoding capacity were then considered and the proposed technique was applied for data decoding purpose. The maximum length of the RFID tag, 8.5 cm, was utilized to encode 17 bits of data while the tag width is only 1 cm, 5.5 times narrower than a credit card size. The effect of aperture size was also considered at this stage. It was found that the actual required aperture size was almost compatible with what was expected through theoretical calculation. The final achieved image resolution was 5 mm in azimuth direction based on the straight trajectory of the reader, the strip map mode.

Finally, the downsides of the proposed technique have been discussed. The requirement for physical movement of the antenna considered as the main limitation. The appropriate solution for this constraint is discussed in the following chapter. The sensitivity of the system to proper tag orientation is also addressed by suggesting the usage of circular polarized antenna instead of linear type.

REFERENCES

1. M. Soumekh, *Synthetic Aperture Radar signal processing with MATLAB algorithms*, John Wiley and Sons, USA, 1999.
2. W. Carrara, R. Goodman, and R. Majewski, *Spotlight Synthetic Aperture Radar Signal Processing Algorithm*. Artesch House, Boston, 1995.
3. G.L. Charvat, *A Low-Power Radar Imaging System*, Michigan State University, 2007.
4. S.K. Yong, P. Xia, and A.V. Garcia, *60 GHz Technology for Gbps WLAN and WPAN: From Theory to Practice*, John Wiley & Sons, United Kingdom, 2010.
5. ACMA, "*60 GHz Band, Millimetre Wave Technology*," 2004.
6. C.A. Balanis, *Antenna Theory Analysis and Design*, Third ed. John Wiley & Sons, Hoboken, NJ, 2005.

7

FAST IMAGING THROUGH MIMO-SAR

7.1 INTRODUCTION

Providing a fine spatial resolution on the chipless RFID tag surface through the millimeter-band 60 GHz and synthetic aperture radar (SAR) technique has been introduced earlier. While the proposed technique benefits from many practical features, there are two main issues that deter its full potentials. The requirement for relative movement of the reader and tag is recognized as the most crucial limitation of the proposed technique. Moreover, the tag orientation sensitivity is also associated with the proposed technique, while a practical solution has been suggested in the previous chapter. This chapter considers the requirement for movement of the reader and a practical solution with reasonable hardware complexity is suggested.

The general structure of the proposed technique based on the conventional SAR technique is shown in Figure 7.1. The reader, including its two orthogonally oriented antennas, moves around the tag and captures the backscattered signal at different view angles. The total synthetic length depends on the reading range, required image resolution, and the tag data capacity. However, as shown in the previous chapters, the maximum length of 30 cm is enough to provide 0.5 image resolution when the tag has its maximum length, 8.5 cm. The interval between each two adjacent transmit and receive antennas also depends on the wavelength. Normally, 2 mm is selected as the physical separation between two transmissions as shown in Figure 7.1.

Advanced Chipless RFID: MIMO-Based Imaging at 60 GHz–ML Detection, First Edition.
Nemai Chandra Karmakar, Mohammad Zomorrodi, and Chamath Divarathne.
© 2016 John Wiley & Sons, Inc. Published 2016 by John Wiley & Sons, Inc.

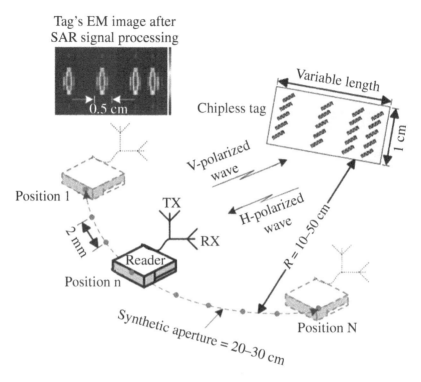

Figure 7.1   System structure of the proposed spatial-based approach.

Therefore, one may conclude that 100–150 individual transmit and receive reading process is normally required. As addressed before, this is a long reading process that is not obviously practical for the current applications of chipless radiofrequency identification (RFID) systems.

This chapter is organized as follows: (i) the phased array antennas with agile beamforming capability are considered as a solution for reader movement and its system complexity is discussed. (ii) The new idea of multiple input multiple output (MIMO)-based array antenna is presented. (iii) MIMO-based array antenna is designed for fast imaging of the tag and the result of simulation is presented. (iv) Finally, the approach for optimization of the MIMO-based system is introduced for minimum number of antennas.

7.2   CONVENTIONAL PHASED ARRAY ANTENNA

Phased array antennas have emerged as the most advanced technology in the modern-day radar systems. It provides huge enhancement in features that are of interest in radar systems, modern wireless communications, and many commercial applications. In a phased array antenna, antenna elements are precisely controlled in relative excitation of phase and amplitude to generate directional beams and nulls in desired direction [1].

Appropriate relative phase adjustments of the array elements in the time domain is known as the time-domain beamforming. This happens by creating precise time delays in the received signals of each array element. Time-domain beamforming provides an accurate result for a wideband signal; however, the precision of the delays is restricted to the capabilities of devices for precise time-delay control. It is also possible to use the fast Fourier transform (FFT) and transfer the required adjustment of the array elements in frequency domain instead of time domain. In this scenario, a proper phase shift in each adjacent element is required instead of time delay. This means that through specific phase shift in the feeding route of each array element, the array can steer on a specific direction. Solid-state digital phase shifters and voltage variable controlled attenuators are widely developed nowadays on which precise adjustment of each antenna element based on the required "weight vectors" is possible. The frequency domain is a more common technique in the phased array systems than the time-delay approach as it can provide much more accurate beamforming results.

Applying the phased array antenna theory to the proposed image-based chipless RFID system, two difficulties should be considered. First, in the phased array structure, phase shifters or their equivalent time-delay modules are needed. To precisely scan the tag surface in the order of a few millimeters, the phase shifters' granularity will be very fine [2,3]. Utilizing accurate phase shifters in the 60 GHz frequency band significantly increases the system cost. Moreover, when the phased array antenna scans the tag, the beam shape distorted with the view angle. This may not be a serious issue in many applications; however in the proposed theory, the fixed shape of the radiation pattern is vital, otherwise, the tag's image is distorted [4].

To minimize the variation of the agile radiation pattern at different view angles, it is possible to use a very accurate feeding algorithm that provides the required amplitudes and phase shifts to individual elements [2]. In this approach, more accurate amplitude tapering modules and precise phase shifters are required that significantly increase the system's cost and complexity [2]. Therefore, phased array antennas with fixed beam shape are normally restricted to very-high-technology military applications. The proposed low-cost EM-imaging chipless RFID system cannot tolerate such an expensive and complex antenna system.

With the aforementioned limitations, it can be concluded that the usage of a conventional phased array antenna for replacement of the moving reader is not practically feasible for the RFID application. Therefore, an alternative to the conventional phased array antenna – a MIMO-based array antenna with reduced complexity – is considered next.

7.3 MIMO-SAR SYSTEMS

MIMO refers to a system with multiple transmit and receive antennas for multiplying the capacity and performance of the radio systems by exploiting the spatial diversity. Considering the potential advantages on providing flexibilities and freedom due to the usage of multiple channels, the MIMO-based systems are of interest to researchers [5]. MIMO technique has recently provided noticeable simplifications on radar systems [6–8]. MIMO-SAR employs multiple antennas to transmit orthogonal waveforms and multiple antennas to receive radar echoes. MIMO-SAR is recently proposed in remote sensing concept [9]. It is shown that the MIMO-SAR can be used to improve the remote sensing system performance. One of the main advantages of the MIMO-SAR is greatly increased degrees of freedom by the concept of virtual antenna array [6]. In a conventional phased array radar system, the transmitted signals are inherently dependant. They only vary by some time delays or equivalent phase shifts. However, in the MIMO-based system, each antenna transmits a unique waveform that is orthogonal to the waveforms transmitted by other antennas. This means that the return signals to each receive antenna element will carry independent information about the target, hence increasing the system's freedom and flexibility. The phase difference caused by different transmitting antennas along with the phase differences caused by different receiving antennas can form a new virtual antenna array steering vector [6]. With optimally designed array elements' positions, a very long array steering vector with a small number of antennas can be created. This provides high flexibility and reconfigurability in antenna configuration [6,10].

The idea of the MIMO-based array antenna is applied on the chipless RFID system. It is found that the new technique of MIMO-SAR system has huge potential for enhancing the tag imaging time on a fairly low-cost and simple reader configuration.

7.3.1 MIMO-Based Array Antenna

An eight-element MIMO-based antenna including four transmit and four receive antennas is shown in Figure 7.2. Each of transmit and receive antennas can be viewed as a sparse array. The normalized separation distance between Rx and Tx antennas are 6 and 9, respectively, as shown in Figure 7.2. Dimension is not important, but centimeter can be assumed as example. Therefore, the sparse array antennas have constant separation of 3 cm. This means that the Rx array is equivalent to a sparse

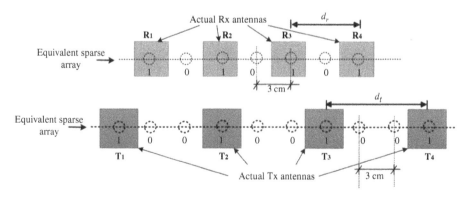

Figure 7.2 MIMO-based antenna system and the equivalent sparse array antennas.

array of seven elements with a separation distance of 3 cm and the active and passive elements distribution of

1	0	1	0	1	0	1

The same assumption is valid for the Tx array with the following topology for the active and passive elements and again with a separation distance of 3 cm:

1	0	0	1	0	0	1	0	0	1

It is envisioned to show the transition between the MIMO theory and its equivalent virtual array antenna without being deeply involved with the mathematical calculations [11]. In Figure 7.2, each Tx antenna can independently transmit its signal and all the Rx receive antennas collect the backscattered signal. Based on the array theory, it can be shown that the array factor of each array antenna is equivalent to a polynomial of $N-1$ order on which N is the number of array elements [12,13]. Therefore, the related polynomials of receive and transmit arrays are as follows, respectively:

$$P_r = 1 \cdot z^6 + 0 \cdot z^5 + 1 \cdot z^4 + 0 \cdot z^3 + 1 \cdot z^2 + 0 \cdot z + 1$$
$$= z^6 + z^4 + z^2 + 1; \quad \text{for Rx array}$$
$$P_t = 1 \cdot z^9 + 0 \cdot z^8 + 0 \cdot z^7 + 1 \cdot z^6 + 0 \cdot z^5 + 0 \cdot z^4 + 1 \cdot z^3 + 0 \cdot z^2 + 0 \cdot z + 1$$
$$= z^9 + z^6 + z^3 + 1; \quad \text{for Tx array} \tag{7.1}$$

It can be mathematically proven that the two sets of Tx and Rx antennas on MIMO-based configuration is equivalent to a virtual array antenna. This means that instead of analyzing the MIMO-based antenna, one may consider the equivalent virtual array and expect almost the same radiation characteristics as the MIMO-based antenna. The details of expressions for relation between the MIMO-based system and the virtual array are not discussed here. It has been shown that the convolution of the Tx and Rx arrays defines their equivalent virtual array antennas and subsequently specifies the position of the active elements in the virtual array configuration [6,11]. Therefore, two sets of the Tx and Rx antennas that have been considered as sparse arrays are mathematically linked to their virtual array antenna. Based on this assumption, the convolution of the Tx and Rx antennas in the time domain is equivalent to multiplication in the frequency domain. Hence, one may simply find the equivalent virtual array's polynomial through the following relation [6]:

$$P_v = P_r \cdot P_t = z^{15} + z^{13} + z^{12} + z^{11} + z^{10} + 2z^9 + z^8 + z^7$$
$$+ 2z^6 + z^5 + z^4 + z^3 + z^2 + 1 \tag{7.2}$$

where P_v is the polynomial of the virtual array. This polynomial that is in the order of 15 relates to a 16-element linear array with the following active/passive topology:

| 1 | 0 | 1 | 1 | 1 | 1 | **1** | 1 | 1 | **1** | 1 | 1 | 1 | 1 | 0 | 1 |

In the above configuration, "0" means that position has no antenna and "1" represents an active antenna. The bold "1" show duplicate element position. With this procedure, the MIMO-based structure of $4 + 4$ physical elements is related to a virtual array of 4×4 elements. The actual MIMO based system and its equivalent virtual array antenna are presented simultaneously in Figure 7.3, in which the physical MIMO-based system and its equivalent virtual array antenna are presented simultaneously.

To have a more physical sense of evolution from MIMO-based system to the array antenna concept, one can track the signal route when it is transmitted by a specific transmit antenna (T_i) and received by all receive antennas. Let consider d_r and d_t as the separation distance between the Rx and Tx antennas, respectively. The relative position of each Tx antenna to the center point of the transmit array antenna is denoted by d_{tn} and the relative position of each Rx antenna to the center point of receive array is shown by d_{rm}, where $m, n = [1, 4]$.

When one of the transmit antennas is active, then the received signals to all receive antennas can be weighted by

$$w_{m,n}(\theta_0) = e^{j\left(\frac{2\pi f}{c}\right)(d_{tn}+d_{rm}).\cos(\theta_0)} \tag{7.3}$$

This clearly shows that the relative positions of each set of transmit and receive antennas and the relative phase shift in its equivalent virtual array. To establish a unique phase center for all communication links between the transmit and receive antennas, one can shift the position of transmitters to a fixed point, the center of transmit array, for instance, and then find the phase shift occurring in all receive array elements. This is shown for the two cases of T_1 and T_2 in Figure 7.4. While T_1 antenna

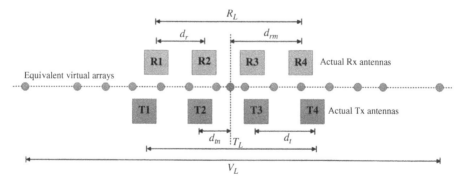

Figure 7.3 MIMO-based antenna and its equivalent virtual array antenna.

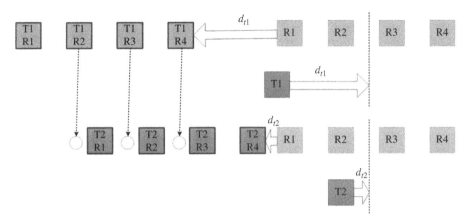

Figure 7.4 Relative phase shift between one particular Tx antenna and all Rx antennas.

is d_{t1} shifted to the right side, all its related Rx antennas are equivalently shifted to the left side for the same distance. The same happens for case T_2. However, the distance of movement is now d_{t2}. If the shifted Rx antennas for these two cases are mapped to each other, they fill the gaps among physical elements. This is shown in Figure 7.4 by the dashed arrows and circles among active antennas. If the same procedure continues for T_3 and T_4, then the virtual array element is completed and would be similar to the declared virtual array polynomial and the position of the antenna in the virtual array as follows.

$$1 \quad 0 \quad 1 \quad 1 \quad 1 \quad 1 \quad 1 \quad 1 \quad 1 \quad 1 \quad 1 \quad 1 \quad 1 \quad 1 \quad 0 \quad 1$$

7.3.2 MIMO-SAR Array for Chipless RFID Imaging

To design the MIMO-based phased array antenna for the proposed chipless RFID system, the process described in Ref. [14] is followed. In the proposed configuration, every two Rx antennas are linked with five Tx antennas. Then the structure is cascaded to provide the larger required equivalent array antenna. Figure 7.5 shows the structure of this MIMO-based antenna configuration. It is important to find the relation between the separation distance of Rx and Tx antennas and their equivalent virtual array. The angles θ_1 and θ_2 can be linked to other parameters as follows:

$$\tan \theta_1 = \frac{2y_0}{X_{r0}}; \qquad \tan \theta_2 = \frac{2y_0}{X_{r0} + \Delta X_r} \qquad (7.4)$$

where the parameters of (7.4) are defined in Figure 7.5. To find the position of the points a and b, one can write the following expressions:

$$x_a = X_{r0} - \frac{y_0}{\tan \theta_1} = X_{r0} - \frac{y_0(X_{r0})}{2y_0} = \frac{1}{2}X_{r0}$$

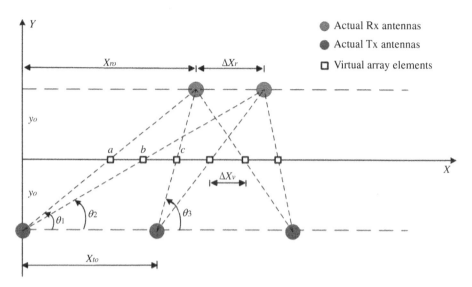

Figure 7.5   MIMO-based antenna and its equivalent virtual array.

$$y_b = X_{r0} + \Delta X_r - \frac{y_0}{\tan \theta_2} = X_{r0} + \Delta X_r - \frac{y_0}{2y_0}(X_{r0} + \Delta X_r) = \frac{1}{2}(X_{r0} + \Delta X_r)$$

$$(7.5)$$

Therefore, the separation distance of Rx antennas can be easily linked to the separation among the virtual array elements, ΔX_v, through

$$x_b - x_a = \Delta X_v = \frac{1}{2}\Delta X_r \tag{7.6}$$

If the same approach is followed for θ_3 and the distance between points b and c is found, one can easily reach the following relation:

$$X_{t0} = 2\Delta X_r = 4\Delta X_v \tag{7.7}$$

We have already discussed that the reader antenna has to illuminate the tag in every 2 mm. Therefore, the inner space of virtual array element shall be 2 mm and equivalently, the physical Rx antennas have to be 4 mm and Tx antennas 8 mm apart from each other.

The total number of sampling points also governs the total number of required physical antennas. We have already discussed that "$m + n$" elements in Rx and Tx configuration may provide a virtual array of "$m \times n$" elements. It is important to mention that this is a theoretical criterion and shows the minimum number of required actual Tx and Rx antennas. For example, 8 and 13 physical antennas are utilized in Ref. [14] for a virtual array with only 44 elements while based on the theoretical relation, less number of antennas are expected for the virtual array with 44 elements.

For the case of RFID application, it was shown earlier that the reader shall stop in 100–150 individual points and illuminates the tag and capture the backscattered signal. Thus, the average number of virtual array elements can be assumed as 125 points; however, it is possible to have slightly higher or even less number of array elements. Therefore, 125 points can be seen as a rough estimate. The proposed MIMO-based system topology for the Rx antenna in Ref. [14] includes two active antennas and then four passive elements in the sparse array configuration, 11000011 array polynomial. The Tx array has one active and one passive element scheme, 101010, for instance. Based on this assumption, the antenna distribution scheme is as follows:

For Tx array antenna: 110000110000110000110000 11
For Rx array antenna: 101010101010101010101010 101

Based on this array distribution, the following polynomials can be related to the Rx and Tx antennas, respectively, for the 44 elements requirements:

$$P_r = Z^{19} + Z^{18} + Z^{13} + Z^{12} + Z^7 + Z^6 + Z + 1 \tag{7.8}$$
$$P_t = Z^{24} + Z^{22} + Z^{20} + \cdots + Z^4 + Z^2 + 1$$

Obviously, the convolution of the above polynomials results in a virtual array of degree 43. If the same topology is followed for the case of 125 array elements, one may suggest the following polynomial for the Rx array antennas including 24 active antennas:

$$P_r = z^{67} + z^{66} + z^{61} + z^{60} + z^{55} + z^{54} + z^{49} + z^{48} + z^{43} + z^{42} + z^{37}$$
$$+ z^{36} + z^{31} + z^{30} + z^{25} + z^{24} + z^{19} + z^{18} + z^{13} + z^{12} + z^7 + z^6 + z + 1 \tag{7.9}$$

Based on the above assumption for the Rx configuration, the related Tx antennas can be defined through the following polynomials with 31 physical Tx antennas:

$$P_t = Z^{60} + Z^{58} + \cdots + Z^4 + Z^2 + 1 = \sum_{k=0}^{30} Z^{2k} \tag{7.10}$$

The convolution of (7.9) and (7.10) results in a polynomial of degree 127, which includes all the required elements.

$$P_{total} = \sum_{i=0}^{127} a_i \cdot Z^i \tag{7.11}$$

In Equation (7.11), coefficients a_i may be different from "1." In this case, it is not required to consider all relevant combinations between an individual Tx and Rx antennas. Therefore, through $(31 + 24)$ physical antenna, it is possible to implement

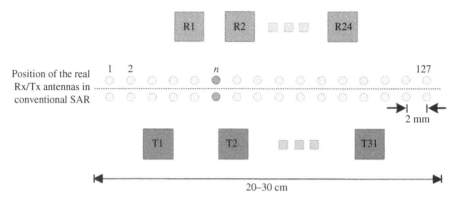

Figure 7.6 Conventional SAR and MIMO-SAR in a glance.

a virtual array with 127 elements. Through this configuration, the total number of array elements drops from $127 \times 2 = 254$ to only $31 + 24 = 55$ elements. Moreover, the reading time in the proposed technique would significantly enhance while no complex system structure is utilized. The conventional SAR technique through physical antennas and its equivalent MIMO-SAR are shown in Figure 7.6. In the conventional SAR, a pair of Tx and Rx shall move around the tag and stop in 127 positions and perform the reading process. Equivalently, 127 pair of antennas can be used for fast reading process. However, in the MIMO-SAR technique, only 55 physical elements are utilized with no requirement for physical movement of antennas.

The summary of the MIMO-based system can be compared with a physical array and the conventional SAR technique as mentioned in Table 7.1. The conventional SAR technique has the minimum system complexity due to the usage of only one antenna pair. However, it provides a very slow imaging process in conventional approach or fairly slow imaging speed if a stepper motor is utilized. It is possible to develop an array antenna for fast imaging process while it requires a large number of elements, which results in significant system cost. Instead, one may use the MIMO-SAR system with much lower number of elements and the requirement for a switching network.

TABLE 7.1 The MIMO-Based Advantages

	Number of Elements	System Complexity	Imaging Time
MIMO-based antenna	$M + N$(min)	Switching network	Fast
Array antenna	$M \times N$	Phase shifter/amplitude tapering	Fast
Conventional SAR	1	Physical antenna movement	Very slow/slow (stepper motor)

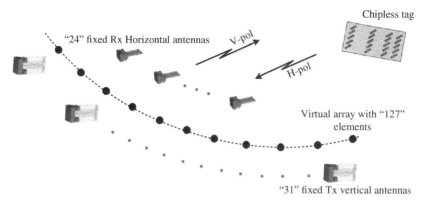

Figure 7.7 MIMO-based antenna for EM-image-based chipless RFID system.

7.3.3 Simulation Result

Figure 7.7 shows the proposed structure of the MIMO-based system. 24 Rx antennas with 4 mm separation distance are linked with 31 Tx antennas with a separation distance of 8 mm. Hence, the total aperture size is less than 25 cm ($31 \times 0.8 = 24.8$ cm) that is completely matched with the aperture sizes introduced before. The virtual array antenna with 127 elements is located between the Rx and Tx antennas. No phase shifter or amplitude tapering network is required. Only a switching network is utilized to connect one Tx antenna with its associated Rx antennas at a time.

To test the validity of the proposed technique of a MIMO-based reader, a 6-bit tag is selected as example. The tag's image is simulated through a MIMO-SAR system and then compared with the image through normal SAR technique. The result of signal processing is shown in Figure 7.8. In the SAR technique that includes 125 individual send and receive process, the EM image adequately shows the encoded data. If the same tag is processed through a MIMO-SAR system that utilizes 55 physical antennas, then the image is depicted in Figure 7.8(c). The EM image seems to be somehow noisier than that for the conventional SAR. Moreover, the imaged column of meander lines shows a minor inclined nature so slightly degrades the azimuth resolution. One reason may be suggested for the degraded image of the tag through MIMO-SAR technique. The equivalence between the virtual antenna and its associated MIMO system is an approximated relation. Each set of Tx and Rx antennas is associated with one single element in the virtual array. However, based on the physical distance of Tx and Rx antennas, a certain amount of error due to the phase center is expected. This error is reflected in the nonperfect image of the tag through MIMO-SAR technique. However, much faster imaging time of MIMO-SAR system compensates the noisier image of the tag. Moreover, it is possible to use higher number of physical antennas on the MIMO-based system and improve the image quality. Alternatively, by optimization of the antennas distribution on the MIMO-basis system, one may expect a higher quality of the imaged tag. The optimization process of the antennas distribution in the MIMO-SAR system is introduced in the following section.

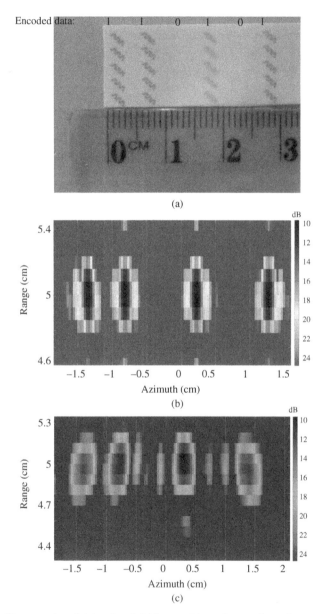

Figure 7.8 Photographs of (a) printed 6-bit tag, (b) tag image through normal SAR, and (c) tag image through MIMO-SAR technique.

7.4 OPTIMIZATION

As discussed in the previous section, the usage of 24 Rx and 31 Tx antennas is based on the suggested topology in Ref. [14] on which 2 Rx and 5 Tx antennas are connected to each other. Then the structure can be expanded by cascading them to provide the required aperture size. This scenario is also based on the linear sparse array theory assumption. Obviously, the number of utilized antennas, 55 elements, is not easy to implement and may put restriction on the practicality of the proposed MIMO-based approach. Moreover, the expectation is to have a virtual array of 127 elements with much lower number of physical Tx and Rx antennas, $11 + 12$ elements, for example. This simply means that the proposed topology in Ref. [14] is not optimum and it would be probably possible to reduce the number of actual antennas and still expect the same virtual array size. To optimize the MIMO-SAR system for the minimum number of physical antennas, two approaches can be considered: analytical and numerical optimization approaches.

7.4.1 Analytical Approach

First, it is suggested to represent the problem analytically and explore if there is any mathematical approach to find the minimum number of physical antennas in the MIMO-SAR configuration. The aforementioned situation for the MIMO-SAR can be analytically shown as

$$
\begin{aligned}
&P_r \cdot P_t = 1 + Z + Z^2 + Z^3 + \cdots + Z: \quad \text{The targeted polynomial} \\
&\deg(P_r) = m \rightarrow \deg(P_t) = n - m; \\
&NoE(P_r) = k; \quad \text{Number of } Rx \text{ element} \\
&NoE(P_t) = t; \quad \text{Number of } Tx \text{ element} \\
&c_1 \leq \frac{k}{t} \leq c_2 \\
&\min(k + t)
\end{aligned}
\tag{7.12}
$$

In the above set of expressions, the product of $P_r \times P_t$ shall produce a complete polynomial of order "n." The number of active elements in the Rx and Tx series is shown by "k" and "t," respectively. There are some other limitations that shall be considered in the optimization process. First, as it is mentioned before, the relation between any Tx and Rx antennas with their equivalent virtual element is an approximate relation. This relation is more accurate if the distance between the Tx and Rx antennas is negligible with respect to the distance of the target in front of the antennas. This implies an additional limitation in the above sets of formula in Equation (7.12). Every Rx antenna shall be connected to its neighboring Tx elements and not to the Tx antennas that are physically located far from the intended Rx antenna. Otherwise, the results of analytical optimization may be mathematically accurate but results in a system with inaccurate performance. Second, the goal is to minimize $k + t$, while it is also desirable to have the minimum physical space occupied by the MIMO-based system. The other parameters of the system shall also be considered in the above sets of formula. Considering the limitations set out for the optimization problem, any analytical approach would be very complex if any existed. The authors are not aware of any analytical approach for solving the above-described problem.

7.4.2 Numerical Optimization Process

The second approach is the numerical optimization process. In the analytical approach, the suggestion is to consider the distribution of the Tx and Rx antennas as a uniform sparse array antenna for simplicity. In the numerical approaches, however, such a suggestion is not mandatory. This means that nonuniform sparse array structures can also be considered, which may result in a lower number of antennas or smaller array size.

Before selection of any specific optimization technique, it is useful to explore the dimension of the problem. This helps selecting the best technique with least complexity. The maximum length of the demanded virtual array for the proposed application, as discussed before, is 30 cm. It is necessary to find out what is the length of Tx and Rx arrays in the MIMO system. Referring to Figure 7.3 and by following the approach of shifting the Rx array for the relative position of each Tx antenna, as described in Section 7.3.1, one may easily prove

$$R_L + T_L = V_L = 30 \text{ cm} \tag{7.13}$$

where R_L, T_L, and V_L are the physical length of Tx, Rx, and virtual array, respectively. To find out R_L and T_L, another equation is required. This means that the relative length of Tx and Rx arrays should be known. Although no specific relation can be defined for the length of Tx and Rx arrays, the ratio of R_L to T_L that has been suggested by Lincoln Laboratory [11] may provide a practical value:

$$R_L \approx \frac{2}{3T_L} \tag{7.14}$$

Considering Equations (7.13) and (7.14), one may easily find the total length of Rx and Tx as

$$\text{Rx} = 12 \text{ cm and Tx} = 18 \text{ cm}$$

The maximum length of the virtual array is 30 cm and it was already shown that the reading process shall be occurred in each 2 mm. Therefore, one may see the virtual array as a uniform array of 150 elements with a uniform distribution and an element spacing of 2 mm. Referring to Figure 7.5, the minimum element spacing of the Rx and Tx arrays shall be 2 mm or lower. For simplicity of optimization, the 2 mm is selected. Consequently, the Rx and Tx arrays on MIMO basis can be viewed as nonuniform sparse array of 60 (12 × 5) and 90 (18 × 5) elements. The total number of physical antennas on Tx and Rx is restricted by

$$k \cdot t \geq 150 \tag{7.15}$$

on which k and t are the total number of real Rx and Tx antennas as already defined in Equation (7.12). A good suggestion for k and t would be 11 and 14, respectively. Therefore, the optimization problem has been simplified as follows.

The total number of 11 and 14 antennas will be distributed appropriately on sparse arrays with 60 and 90 elements, respectively, on which two arrays are connected on MIMO basis. The distribution of antennas on each array will be in such a way to result in a complete virtual array of 150 elements. Therefore, the total possible option for each case is as follows:

$$\text{Possible options for Rx array:} \quad \binom{60}{11} = 1.3 \times 10^{19}$$
$$\text{Possible options for Tx array:} \quad \binom{90}{14} = 7.88 \times 10^{26} \tag{7.16}$$

Therefore, the total number of possible cases would be more than 10^{45}. This shows very wide dimension of the answer region for the suggested problem. Obviously, it would not be possible to verify all the possible cases even with the most advanced computer systems. Instead, an efficient optimization process is required to find the solution in a reasonable time frame.

Luckily, this is a common issue nowadays in many research fields, problems with widely extended feasible region without any known classical approaches. Applying the numerical techniques seems to be the only possible method. Moreover, classical numerical methods fail to provide solution in many cases as the target function does not have specific characteristics, being a continuous function or having derivatives, for example. In such cases, global optimization approaches are usually the best option. Another benefit of global optimization approach is their ability on finding the global optimum of a function. One of the main downsides of the classical optimization techniques is their failure on finding the best possible solution for a certain problem as they can easily be entrapped in local minima. Moreover, these techniques cannot generate or even use the global information needed to find the global minimum for a function with multiple local minima. The results of classical optimization approaches normally depend on the initial suggestion or starting point.

Mentioning the drawback of the global optimization techniques, one may refer to their computationally costly process due to the slow convergence. This means that irrespective of their significant ability of directing the optimization process toward the global optimum solution, they are normally inefficient to find the final answer. One of the main reasons for their slow convergence is that they may fail to detect promising search directions especially in the vicinity of local minima due to their random constructions. Combining global optimizations with local/classical search methods is a practical remedy to overcome the drawbacks of slow convergence and random constructions of global optimization techniques, hybrid methods. The advantages of global approaches are used to direct the process to the vicinity of global optimum and a classical technique is normally added at the final stage of the search.

In the global optimization techniques, evolutionary algorithms (EAs) are an important subset that generate solutions to optimization problems using techniques inspired by natural evolution. There are many techniques in EA category in which the genetic algorithm (GA) is one of the most significant approaches with the most effective results for a wide range of problems. GA is specifically very effective if no real-time solution is required [2,15]. GA is proposed for the current problem of MIMO-SAR technique. The rest of this section briefly introduces GA and some of its main functions. Readers are suggested to refer to special references on GA for more detailed information.

7.4.3 Genetic Algorithm

Genetic algorithm that is been inspired by natural evolution encodes a potential solution to a specific problem on a simple chromosome-like data structure and applies recombination operators to these structures so as to preserve critical information. There are some parameters and key components in GA that will be defined and introduced. The most important ones are as follows:

Genes: Genes are the essential blocks of the GA that carry the genetic information of an individual. Normally, genes represent the coded version of the physical parameters.

Chromosome (genotype): A chromosome is a certain set of genes with a known length that define a proposed solution to the problem that the GA is aimed to optimize.

Population: A group of chromosomes is called population.

Generation: GA works on a specific population at each specific time that is called generation.

Parents: Two or more chromosomes from the current population that have been selected to transfer their genetic characteristic to the next generation.

Children: Referring to a chromosome belonging to the next generation and is the result of crossover among parents in the current population.

Fitness: It is a number that is assigned to a chromosome and shows how well the chromosome satisfies the requirement of the problem.

There are four main operators in GA that control the flow of optimization process. First is the transformation and mapping of the real problem to the chromosome environment. The chromosome space is normally a binary-type data that shows the parameters and variables of the real problem. Then a certain number of chromosomes are selected to create the current population. After the construction of a population, parents have to be nominated for getting the chance of preserving their critical information. Normally, chromosomes with the highest fitness will have more chance of being selected as parent. However, this does not mean that the weaker individuals should have no chance for passing their genes to the next generation. Parents then do crossover and create children that belong to the next generation. There is the possibility of mutation in the children genes as well. This process will continue until one

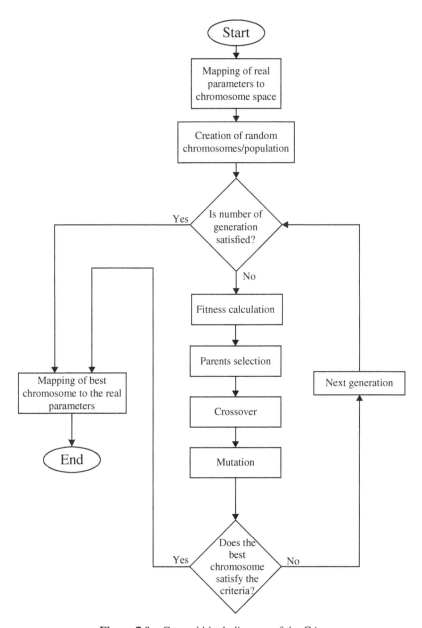

Figure 7.9   General block diagram of the GA.

chromosome satisfies the requirement of the problem or a certain number of generation is passed. The main block diagram of the GA is shown in Figure 7.9. Readers may refer to the special references [15,16] for better and more detailed explanations about each operator of the GA.

7.4.3.1 Chromosome and Population Structure for MIMO-SAR As discussed before, Rx and Tx arrays are assumed as sparse arrays with minimum interelement of 2 mm and maximum length of 12 and 18 cm, respectively. Each of Rx and Tx is sparse array with 60 and 90 elements, respectively, which includes certain number of physical antennas. Therefore, each chromosome shows one possible distribution of the physical antennas on the Rx and Tx arrays. The dimension of each chromosome is selected to be a matrix of [1,150]. The first 60 elements show how the physical antennas are distributed on Rx array and the rest, elements 61–150, show the scattering of physical antennas on Tx sparse array. A sample of chromosome is shown in Figure 7.10.

In this figure, the first 60 numbers relate to Rx array and show how the physical antennas are distributed. "1" means that position in the sparse array contains an active/physical antenna and "0" shows a blank point on the sparse configuration. As one may notice, there are only 11 physical antennas on the Rx section and the rest are zeros. The same is true for the Tx array that is shown for elements 61–150 with 14 "1" showing the physical antennas.

The next step is to create the first population. If the size of population is selected to be 50, then 50 random chromosomes similar to the structure shown in Figure 7.10 will be created. Therefore, the population has the dimension of 50×150 as shown in Figure 7.11.

7.4.3.2 Fitness Calculation Each individual/chromosome in the population suggests a certain arrangement of the physical antennas for the Rx and Tx array

0	0	1	0	0	1	0	0	1	0	0	0	0	0	0	1	0	0	0	0
1	0	0	0	0	1	0	0	0	0	0	0	0	0	0	1	0	0	0	0
1	0	0	0	0	0	0	0	0	0	0	0	0	1	0	0	1	0	1	0
1	0	0	0	0	0	0	1	0	0	0	1	0	0	0	0	0	0	0	0
0	0	0	1	0	0	0	0	0	0	1	0	0	0	0	0	0	0	0	0
1	0	0	1	0	0	0	0	0	0	0	0	0	1	0	0	1	0	0	0
0	0	1	0	0	1	0	0	0	0	0	0	0	0	0	0	0	1	1	0
0	0	0	0	0	0	0	1	0	0										

Figure 7.10 Sample chromosome for the MIMO-SAR optimization, red values relate to Rx and black numbers show the Tx array.

$$\text{Population} = \begin{bmatrix} 0 & 1 & 0 & 0 & 0 & 1 & \dots & 0 & 01 & 0 \\ 1 & 1 & 0 & 1 & 0 & \dots & 0 & 0 & 1 & 0 & 0 \\ & & & \vdots & & & & \\ 0 & 0 & 0 & 0 & 1 & \dots & 1 & 0 & 0 & 0 & 1 \end{bmatrix}$$

$$50 \times 150$$

Figure 7.11 Structure of population in GA.

structures. It is required to evaluate and quantify the suitability of each individual by assigning a value as the fitness to each chromosome.

For calculation of the chromosome's fitness, the related polynomial for each of Rx and Tx arrays will be calculated. As an example, the related Rx and Tx polynomials for the chromosome of Figure 7.10 are as follows:

$$Rx \text{ polynomial}: x^{57} + x^{54} + x^{51} + x^{44} + x^{39} + x^{34} + x^{24} + x^{19} + x^{6} + x^{3} + x$$
$$Tx \text{ polynomial}: x^{89} + x^{82} + x^{78} + x^{66} + x^{59} + x^{49} + x^{46} + x^{36} + x^{33}$$
$$+ x^{27} + x^{24} + x^{12} + x^{11} + x^{2} \qquad (7.17)$$

As mentioned before, the virtual array is the result of Rx and Tx convolution in the time domain or multiplication of their polynomials in the frequency domain. Therefore, two polynomials of Equation (7.17) are multiplied to result in:

$$x^{146} + x^{143} + x^{140} + \cdots + x^{12} + x^{8} + x^{5} + x^{3} \qquad (7.18)$$

In an ideal scenario, Equation (7.18) is a complete polynomial of order 149 without any zero coefficient. However, as the sample chromosome that resulted in Equation (7.18) is not the best individual, hence Equation (7.18) includes some zero coefficients.

Based on this, one may define the fitness in such a way to show the number of zero coefficients of the resulting polynomial for each chromosome. Therefore, the lower is the fitness means less number of zero coefficients and hence the better the chromosome.

After calculating the fitness of all individuals in the initial population, the chromosomes in the population are rearranged based on their fitness. The best chromosome, which has the lowest fitness value, goes to the top and the worst chromosome with highest fitness value will be located at the end of the population. This is required for selection operator, which is discussed in the following section.

7.4.3.3 Selection Operator   After arranging the population based on the chromosomes' fitness, some individuals are to be selected as parents. Obviously, chromosomes selected as parents will have the chance of transferring their genes to the next generation. The nonselected chromosomes will die and their genetic information no longer existed in the optimization process. Therefore, it is important to make sure that the algorithm provides a rational process for preserving the critical genetic information of all chromosomes by giving them enough chance of being selected as parents based on their fitness values. There are many suggested approaches for the selection operator, of which the most common approaches are introduced briefly in the following text.

Population Decimation. This is the simplest strategy for selection of parents. After arranging the population based on the fitness, a random value for fitness is selected. All the individuals who have the fitness worse than the threshold will be deleted. The parents are selected randomly from the remaining individuals. Irrespective of the easy implementation process, this selection strategy deletes all the weak individuals without giving them any chance of participating in the formation of the next generation.

Proportionate Selection. This is one of the most commonly used techniques for the selection operator. Sometimes, it is also referred to as *roulette wheel* approach. In this approach, the chance of selection of each individual relates to its fitness by

$$P_{selection} = \frac{f(parent_i)}{\sum_i f(parent_i)} \tag{7.19}$$

which $f_{parents}$ shows the fitness of ith parents. The implementation of this approach is similar to a roulette wheel on which each chromosome occupies a certain area on the wheel surface based on its fitness as shown in Figure 7.12. The proportionate technique is efficient for the case when the number of population is high enough. Otherwise, some errors are expected for low population sizes. Moreover, the weak chromosomes have a small chance of transferring their genetic characteristics to the next generation.

Tournament Selection. The tournament approach is normally believed to be the most common technique for selection operators. First, a group of chromosomes are randomly selected from the current population. Then the best individual in the selected bunch is the winner and hence will be selected as parent. The selected group will be returned to the population and the process repeats again. Tournament technique provides a satisfactory GA performance and has simple implementation process.

Variable Window Width. The variable window is one of the most effective techniques for the selection operator [15]. After rearranging the population based on their fitness, a window with variable length is applied to the population. Hence, the best chromosome is located at the beginning of the window and the end of the window includes the weakest chromosome in the selected window. Then, one individual is randomly selected from the chromosomes in the window as parent. The function that controls the window width is very important in this

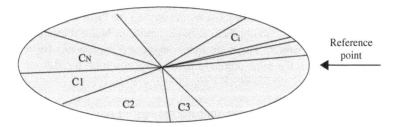

Figure 7.12   Selection operator based on roulette wheel approach.

approach, and the total performance of the algorithm highly depends on this function [2]. The formula that has been suggested in Ref. [15] is as follows:

$$W(L) = \text{Int}\left[2 + \text{Pop}\left(1 - e^{-tL}\right)\right] \tag{7.20}$$

where $W(L)$ is the window width, Int shows the integer value of the parameter, Pop is the total population number, L is a random value that controls the window width, and t is a parameter that has to be selected in such a way to create $W_{max} = \text{Pop}$. Implementing the abovementioned procedure for the selection process, however, did not result in satisfactory operation of the algorithm. Investigating the proposed formula for the selection operator reveals the reason for this unsuccessful performance. If Equation (7.20) is plotted as shown in Figure 7.13, one may notice the reason. For most of the cases ($L > 0.2$), the first 25 chromosomes are having the same chance of being selected as parent. The first chromosome and the 25th individual normally have noticeable fitness value differences. This is almost inconsistent with the overall expectation from selection operator.

Therefore, a new formula is suggested that shows better performance for the GA. The new function is Equation (7.21) and has two parts. When L, which is a random variable, is less than a certain value, 0.3, for example, the window width linearly depends on L. For $L > 0.3$, an exponential function has been suggested.

$$W(L) = \begin{cases} \text{Int}\left(\text{Pop} \times L\right), & L < 0.3 \\ \text{Int}\left(\dfrac{\text{Pop}}{5.58} \times 5.58^L\right), & L \geq 0.3 \end{cases} \tag{7.21}$$

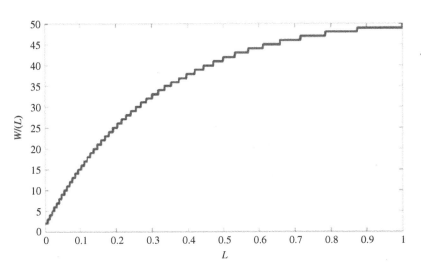

Figure 7.13 Suggested formula by Rahmat-Samii and Michielssen [15] for selection operator.

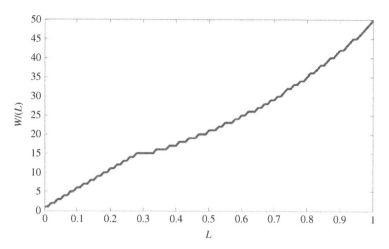

Figure 7.14   Suggested formula for selection operator with best result.

The above-suggested function for the window width is depicted in Figure 7.14. As one may easily see, while the better chromosomes, the first 30% of population, are having a higher chance in the selection process, the weaker chromosomes also have the opportunity to transfer their genetic information to the next generation. It is, however, important to mention that the suggested function in Equation (7.21) is only based on different functions and evaluation of their performances. This simply means that it is possible to find another function that works better for another optimization problem.

7.4.3.4 Crossover   When two chromosomes are selected as parents, they will do crossover, exchanging their genetic information and creating children. Again there are many approaches for doing crossover. The technique that is being used in the proposed application is called "multipoint crossover" technique. In this approach, four random numbers are selected between 1 and 60 and four numbers in the range 61–150. Then each of the parents changes their chromosomes between these breaking points. The crossover operator based on multipoint is figuratively shown in Figure 7.15. For the specific case of the MIMO-SAR, one major issue normally happens during the crossover that should be considered carefully; otherwise, the final result of optimization is completely wrong. As one may notice, the number of "1" in the Rx and Tx arrays is very important as it shows the physical antennas in each array. The important issue is therefore to make sure that when crossover is happening between two parents, the number of "1" in each section, 1–60 and 61–150, should remain unchanged. As an example, if two "1s" are in the genes (R_1, R_2) of Parent 1 and Parent 2 has only one "1" in its genes between R_1 and R_2, then each of the offspring, Child 1 and Child 2, will have wrong number of 1s in their Rx array section after multipoint crossover. Therefore, instead of simply changing the genes of R_1 and R_2 between Parent 1 and Parent 2, the adjustment of 1s should be maintained. For the suggested example, Child

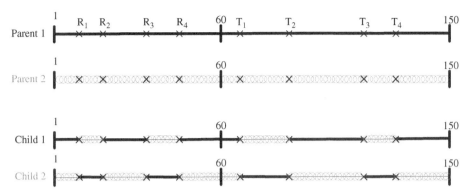

Figure 7.15 Multipoints crossover operator.

1 will have lower number of "1s" in its Rx array. It is possible to add the required number of 1s in its genes (R_1, R_2). On the other hand, Child 2 acquires more 1s in its Rx array section. Hence, one of the 1s will be randomly deleted. The accommodation of this process in the algorithm will be carefully considered.

7.4.3.5 Mutation In GA, mutation is defined as the insertion of random genes to each chromosome. Considering the binary nature of the genes, random genes means changing from "0" to "1" and vice versa. Happening of mutation obviously changes the position of the chromosome on the solution area without any particular trend. Although this may be viewed negatively as it affects the divergence of the operator, mutation is one of the important operators of the GA. Mutation ensures that the GA algorithm does not stick at the local optimums. One may recall that this issue was the main advantages of the global optimization algorithms over classical approaches. At the beginning of the GA algorithm, as the chromosomes are normally different from each other due to the random creation of population, mutation is not of significant importance. However, after enough generations, genes of the chromosomes are closely related to each other because of multiple crossover operators. At this stage, mutation gets major importance as it broadens the search area of the algorithm and prevents it on remaining at the local optimums.

Mutation normally happens in genes of chromosomes by a specific probability. Obviously, the mutation possibility will be remained low at the beginning of the algorithm for minimum effect on the algorithm divergence. When the algorithm proceeds, the possibility of mutation increases broadening the search area of the algorithm. Based on this, the mutation probability starts from 0.08 at the beginning of the algorithm and ends up to 0.5.

Similar to crossover, the number of "1s" in each section of the chromosome is important. Therefore, when one gene of the chromosome encounters mutation, this means that another gene in the same section of the chromosome shall be changed appropriately maintaining the total number of "1s." Therefore, this means that for the specific problem of MIMO-SAR system, the number of mutation in each section, Rx and Tx, is an even value.

7.4.4 Results

Based on the described GA algorithm in the previous section, the MIMO-SAR system is optimized aiming for minimum number of physical antennas in the Rx and Tx arrays, while the resulting virtual array is a complete array of 150 elements. It was already shown that 25 $(11 + 14)$ elements is the minimum theoretical number of physical antennas for the virtual array of 150 elements based on the discussion in Section 7.3.

The written MATLAB codes are able to calculate any number of Rx and Tx arrays. The codes are run for 11 Rx and 14 Tx arrays as already considered in the discussion. The final result of the MIMO system is shown as follows. It is important to mention that few changes and extra steps are also added in the coding process to shift any zero coefficients in the beginning (or ending) of the virtual array. This means that analytical errors happened during the optimization will have less practical effects for the physical arrays.

7.4.4.1 Example 1   In this example, the result of optimization suggests a structure for the Rx and Tx arrays that results in a virtual array of 150 elements with 10 zero coefficients. The suggested Rx and Tx polynomials are as follows:

$$\text{Rx polynomial}: x^{59} + x^{56} + x^{52} + x^{45} + x^{38} + x^{31} + x^{24} + x^{17} + x^{10} + x^{6} + x^{3}$$
$$\text{Tx polynomial}: x^{89} + x^{88} + x^{87} + x^{79} + x^{78} + x^{77} + x^{69} + x^{19} + x^{18} + x^{17}$$
$$+ x^{9} + x^{8} + x^{7} + x^{6} \tag{7.22}$$

The virtual array is shown in Figure 7.16. One may notice, 10 zeros are found in the resulted virtual array coefficient on which 9 of them located at one end of the vector. This zeros have no effect on the array performance except reducing the length of the virtual array from 30 to 28.2 cm. There is only one missing coefficient, A_{142}, which affects the performance of the MIMO system for the SAR purpose. The coefficients that are "2" can be simply converted into "1" by not considering one of the products in the switching network that created that coefficient.

Figure 7.16   Final virtual array with 28.2 cm length and one missing element.

Figure 7.17 Final virtual array with 28.6 cm length and two missing elements.

7.4.4.2 Example 2 The second example covers the same optimization problem; 11 Rx and 14 Tx arrays for a virtual array with 150 elements. The revealed result is as follows:

Rx polynomial : $x^{59} + x^{58} + x^{45} + x^{44} + x^{41} + x^{30} + x^{29} + x^{16} + x^{15} + x^2 + x$

Tx polynomial : $x^{88} + x^{86} + x^{84} + x^{82} + x^{80} + x^{78} + x^{76} + x^{15} + x^{13} + x^{11}$

$$+ x^9 + x^7 + x^5 + x^3 \tag{7.23}$$

Based on the suggested configuration for each of the real arrays, the resulting virtual array is shown in Figure 7.17. The resulting virtual array has seven zero coefficients out of which five are forced at each end of the array. This simply means that the virtual array length is 28.6 cm. Therefore, it is almost long enough to decode the maximum length of the tags with 17 bits of data (see Figure 6.11). However, there are two positions in the virtual array with zero coefficients that may affect the MIMO system performance. Again, the coefficients "2" will be deleted in the system implementation.

7.5 MIMO-SAR RESULTS

The proposed topology suggested by the GA algorithm is applied to the printed tags with different encoding capacity. The results of the tag's content after SAR and MIMO-SAR-based signal processing are presented here.

First, the 6-bit tag shown in Figure 7.8 is simulated based on the optimized MIMO-SAR structure in Section 7.4.4. In the results shown in Figure 7.18, the length of Rx and Tx arrays are 12 and 18 cm, respectively, and hence they resulted in a virtual array with a length of 28.2–28.6 cm. Each of Rx and Tx arrays is designed as a nonuniform sparse array antenna basis while only 11 and 14 active elements are used for Rx and Tx arrays, respectively. Comparing the decoded content by the conventional SAR and through MIMO-SAR technique, one may notice very limited differences. The signal by the MIMO system has higher sidelobe compared with normal SAR technique. Apart from this, two scenarios provide almost the same results. The optimized MIMO-SAR system adequately decodes the tag's data based on the shown EM images of the tag. Both scenarios discussed in Section 7.4.4, Examples 1 and 2, provide almost the same result, hence only one of the results is shown in Figure 7.18.

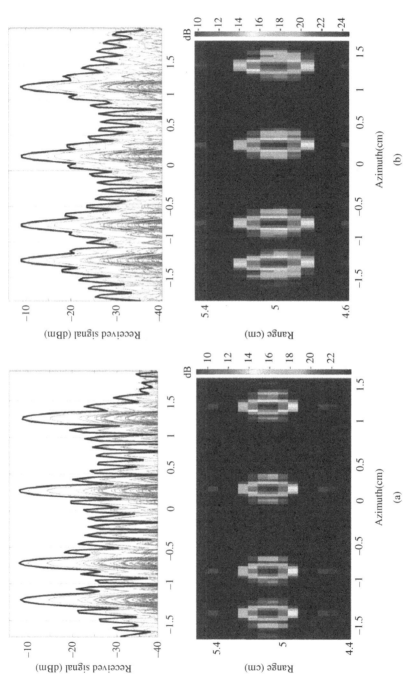

Figure 7.18 Simulated received signal and the tag's image for (a) conventional SAR technique and (b) optimized MIMO system with 25 elements.

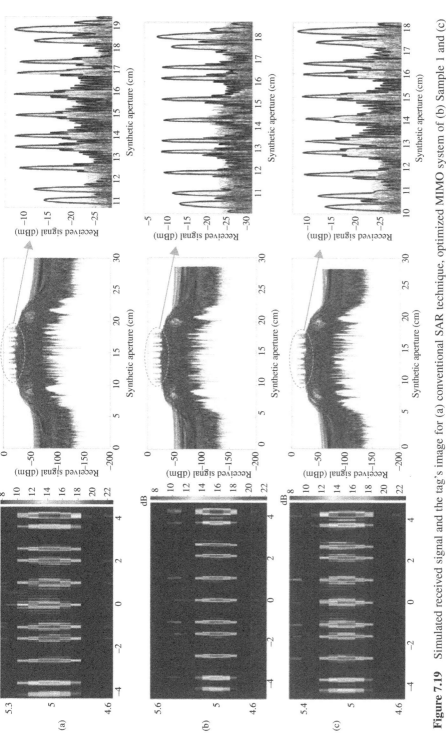

Figure 7.19 Simulated received signal and the tag's image for (a) conventional SAR technique, optimized MIMO system of (b) Sample 1 and (c) Sample 2.

It is important to mention that two results shown in Figures 7.18 and 7.8 are based on different assumptions. The simulated received signal in Figure 7.8 is based on 55 physical elements in a nonoptimized MIMO configuration that results in a virtual array of 125 elements with 25 cm length. On contrary, 25 physical elements in Figure 7.18 are optimized to create a virtual array with 142 elements with 28.6 cm length, as shown in Example 2 of Section 7.4.4.2. Therefore, irrespective of the lower number of physical antennas for the optimized MIMO-SAR, the resulting virtual array has a longer length and more view angles on the tag surface hence resulted in better tag's image.

The same approach is followed for the tag with the maximum data encoding capacity, 17-bit tag with 8.5 cm length as shown in Figure 6.11(b). The results of simulation based on conventional SAR technique, the optimized MIMO of Sample 1 and Sample 2 are shown in Figure 7.19. As it is clear from the figures, all three results are very similar to each other while the implementation process is completely different. The conventional SAR technique requires multiple send and receive at different view angles, which results in very long data capturing process. On contrary, the optimized MIMO-SAR provides a much faster reading process with minimum hardware complexity as it only requires 1 switching network and 25 physical antennas.

Two results in Figure 7.19(b) and (c) are very similar with minor differences in their sidelobe. This difference is only visible if highlighted signal level is viewed; otherwise, two received signals are completely the same. Figure 7.19(b) is based on 28.2 cm of the synthetic aperture length and only one missed array element in the virtual array structure, Example 1 of Section 7.4.4.1. Figure 7.19(c) has an aperture length of 28.6 cm with two missed virtual array elements, Example 2 of Section 7.4.4.2. Both systems use 25 physical antennas.

7.6   CONCLUSION

It was shown that the conventional SAR technique that requires the physical movement of the reader antenna and the tag can be replaced by the MIMO system. The new approach does not require any physical movement instead, a switching network is mandatory. The limitation of the MIMO system is its requirement of considerable number of antennas on its Rx and Tx arrays if the normal procedure on open literature followed. To adequately replace the conventional SAR technique, with 25 cm aperture size, by its equivalent MIMO system, 55 physical antennas are needed. This number of antennas may restrict the practicality of the proposed system. Therefore, the MIMO system was analytically modeled. It was shown that the analytical model is very complex; hence, the related mathematical approach would be very complex if existed. Then the global optimization techniques are suggested to reduce the number of physical antennas without affecting the equivalent virtual array resulted from the MIMO system. Genetic algorithm approach has been applied to find the minimum number of antennas for the MIMO-SAR system. Many techniques have been suggested in the GA for better performance for the proposed RFID application. It was analytically shown that such an approach can successfully optimize the number of physical antennas. The proposed topology by the GA then applied to multiple printed tags structure for decoding purpose. The result of simulation based on the new proposed MIMO structure confirms the performance of the proposed MIMO-SAR technique after optimization process. This means that a MIMO system with minimum number of physical antennas that only requires a switching network may successfully replace the conventional SAR technique. Therefore, there is no requirement for physical movement of the reader antennas around the tag. This new MIMO-SAR system suggests a very fast imaging process with reasonable hardware complexity.

REFERENCES

1. C.A. Balanis, *Antenna Theory Analysis and Design*, Third ed. John Wiley & Sons, Hoboken, NJ, 2005.
2. M. Zomorrodi, "Improved Genetic Algorithm approach for Phased Array Radar Design," in *The Asia-Pacific Microwave Conference, APMC*, Melbourne, Australia, 2011, pp. 1850–1854.
3. Y. Yu, P.G. M. Baltus, A. Graauw, E. der Heijden, and C.S. Vaucher, "A 60 GHz Phase Shifter Integrated With LNA and PA in 65 nm CMOS for Phased Array Systems," *IEEE Journal of Solid-State Circuits,* vol. 45, 2010.
4. M. Soumekh, *Synthetic Aperture Radar Signal processing with MATLAB Algorithms.* John Wiley & Sons, USA, 1999.
5. J. Li and P. Stoica, *MIMO Radar Signal Processing*, John Wiley & Sons, NY, USA, 2008.
6. W. Q. Wang, "Virtual Antenna Array Analysis for MIMO Synthetic Aperture Radars,", *International Journal of Antennas and Propagation, Hindawi Publishing Corporation,* vol. 2012, p. 10, 2012.
7. A. M. Haimovich, R. S. Blum, and L. J. Cimini, "MIMO Radar with Widely Separated Antennas," *IEEE Signal Processing Magazine,* vol. 25, pp. 116-129, 2008.
8. X. Zhuge and A. G. Yarovoy, "A Sparse Aperture MIMO-SAR Based UWB Imaging System for Concealed Weapon Detection," *IEEE Transactions on Geoscience and Remote Sensing,* vol. 49, pp. 509-518, 2011.
9. C. Bill, "Efficient Spotlight SAR MIMO Linear Collection Configurations," *IEEE Journal of Selected Topics in Signal Processing,* vol. 4, p. 7, 2010.
10. D. R. Fuhrmann, J. P. Browning, and M. Rangaswamy, "Signaling Strategies for the Hybrid MIMO Phased-Array Radar," *IEEE Journal on Selected Topics in Signal Processing,* vol. 4, pp. 66–78, 2010.
11. Lincoln Laboratory, "*Practical Phased Array Antenna*," Massachusetts Institute of Technology (MIT), Boston,2014.
12. W.L. Stutzman and G.A. Thiele, *Antenna Theory and Design*, John Wiley & Sons, 1981.
13. M. Zomorrodi, "A Phased Array Antenna Design with Genetic Algorithm", MSc, Electrical Engineering School, K.N. Toosi, Tehran, Iran, 2000.
14. G.L. Charvat, *A Low-Power Radar Imaging System*, Michigan State University, 2007.
15. Y. Rahmat-Samii and E. Michielssen, *Electromagnetic Optimization by Genetic Algorithms*, Wiley, 1999.
16. M. Mitchel, *An Introduction to Genetic Algorithms.* A Bradford Book, USA, 2003.

PART II

ADVANCED TAG DETECTION TECHNIQUES FOR CHIPLESS RFID SYSTEMS

8

INTRODUCTION

Item tagging and monitoring have become significant than ever before, due to the recent emergence of new technologies and their mass-market penetration. Particularly in mass production sites, automated item tagging can increase the productivity and efficiency, which in turn will increase the company revenue. Optical barcode technology is dominant in the item tagging market at present due to its relatively low implementation cost. However, there are a number of limitations such as low reading range, line of sight (LOS) requirement for reading, and its inability to identify multiple items simultaneously. All these challenges suggest barcodes is not a feasible solution in automating item tagging. Radio frequency identification (RFID) shades the light to overcome these limitations toward automating the process, however with high-priced tags. The focus of this research is to produce tags that are comparable to optical barcodes in price, while providing the functionality. This is achieved by removing the microchip and using special techniques hence, called chipless RFID tags. This section of the book presents a few advanced tag detection techniques and a high-data-capacity chipless tag using likelihood-based detection techniques and multiple input multiple output (MIMO) tag design.

8.1 RFID SYSTEMS

RFID is a wireless technology used to automatically identify objects attached to its tags. RFID technology appears to offer as an alternative to optical barcodes due to its unique advantages such as larger reading range, non-LOS reading, multiple tag detection, and its ability to be able to automate the item identification process. A typical

Advanced Chipless RFID: MIMO-Based Imaging at 60 GHz–ML Detection, First Edition.
Nemai Chandra Karmakar, Mohammad Zomorrodi, and Chamath Divarathne.
© 2016 John Wiley & Sons, Inc. Published 2016 by John Wiley & Sons, Inc.

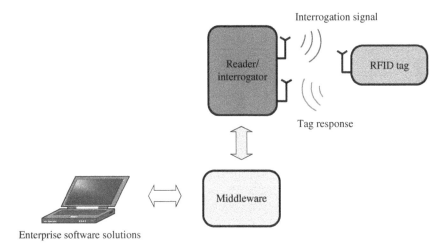

Figure 8.1   A typical RFID system.

RFID system has a reader, a tag, and middleware [1] as shown in Figure 8.1. The RFID reader sends an interrogating signal, which is an electromagnetic (EM) signal, toward the tag and the tag responds back with embedded information to the reader's receiver. Then the reader extracts the information originally encoded by the tag and utilizes the middleware to interface that with the user in a meaningful manner. RFID systems have applications in a number of areas including inventory control, security, logistics, and item tracking.

The RFID reader usually transmits an interrogating signal toward the RFID tag. The tag then modulates the signal with its ID codes and retransmits or backscatters the modulated signal toward the reader. This process is called *tag modulation*. Different modulation techniques are discussed later in the book. The response from the tag is then analyzed to detect and uniquely identify the tag using the signal processing capabilities at the reader. Finally, the middleware integrates the tag identification data with an enterprise software to facilitate the automation process [2].

A vast majority of commercially available RFID tags use application-specific integrated circuits (ASICs) to encode and transmit data. This microchip in the RFID tag makes the tag manufacturing process complicated and expensive compared to optical barcode printing. Researchers have proposed the idea of removing the microchip and using new data encoding techniques. These chipless RFID tags can be printed on paper, read non-LOS, fabricated at low cost, and made fully passive without requiring any energy source [3].

A number of RFID systems have been reported in the literature [1,3–35], which are based on either chipped or chipless RFID tags. In chipped RFID systems, a reasonable amount of processing is done at the tag with the help of a microchip. However, in chipless RFID systems, the tag is given a minimum, if not nil, processing capabilities, as the RFID reader takes all the burden of signal processing. A brief summary of evolution of both single input single output (SISO) and MIMO-based RFID systems is presented next.

8.1.1 SISO-Based RFID Systems

Early research on RFID systems was mainly based on systems having a single antenna at the tag, and also a single antenna at the reader hence named as SISO systems. Since separate antennas for transmission and reception at the reader as well as at the tag enhance the system performance [36,37], researchers proposed multiple antennas at the RFID tags and/or at the RFID readers. It is noteworthy that there is only one dedicated transmission antenna or receiving antenna. These SISO systems can be seen in both chipped and chipless RFID tags proposed in the literature.

However, chipless RFID technologies have still not been able to replace optical barcoding or chipped RFID tags mainly due to lower data capacity, reading range, and tag reading accuracy. Over the last decade, researchers have mainly focused on overcoming these challenges by improving the chipless RFID tag design and the RFID reader architecture [2,3,38–41]. However, they were mostly using primitive signal processing techniques such as moving average or threshold-based detection [42–44]. The few advanced signal processing techniques [45–50] reported so far have high computation complexity; hence, they are not feasible for commercial implementation. It was identified that there is a significant research gap for computationally feasible smart tag detection techniques for chipless RFID systems.

This section of the book addresses the identified research gap and hypothesizes likelihood-based detection techniques for accurate tag detection. The techniques also improve data capacity by removing the guard bands between resonant frequencies of the frequency-domain chipless RFID tags. They also increase the read range by requiring low signal-to-noise ratio (SNR) in detected signals. Then the tag reading was made faster with computationally feasible tag detection techniques such as trellis-tree-based Viterbi decoding and bit-by-bit suboptimal approaches.

8.1.2 MIMO-Based RFID Systems

The rapid development of RFID devices and their wide use in mass market applications have prompted researchers to work mainly on improving RFID system performances. The performance metrics include interrogation range, bit-error-rate (BER), data rate, anticollision, and implementation cost. In wireless communication, most of these performance metrics have been improved using MIMO antenna technology. As a result, during the last few years, there has been a significant research focus on applying MIMO technology to chipped RFID systems. The most important MIMO-based chipped RFID systems reported in the literature are revisited here.

Chizhik *et al.* [51] introduced the concept of pinhole/keyhole when describing the capacities of multielement transmit and receive antennas. The formulation of a pinhole can be visualized in the following example. Picture a two-element transmitting array and a two-element receiving array that are separated by a screen with a small keyhole/pinhole punched through it. The only way for the radio wave to propagate is to pass through the pinhole. A hallway or a tunnel is perhaps a more realistic environment where the pinhole concept can be experienced. In Ref. [36], Griffin has shown that pinhole diversity is available in a rich scattering environment caused by the modulating backscatter with multiple RF-tag antennas.

In Ref. [37], a ultrahigh frequency (UHF)-chipped RFID system has been investigated with multiple readers, where the channel from transmitter to the receiver of the reader, via the RFID tag, was assumed to be a pinhole channel. Both forward (from the reader to tag) and reverse (from the tag to reader) links are assumed to be having a *Nakagami-m* fading channel [52]. An $M_t \times 1 \times M_r$ pinhole channel has been investigated further, where there are M_t transmitting reader antennas, one RF-tag antenna, and M_r receiving antennas at the reader. There are two system configurations analyzed in this work, namely, *monostatic system* with transmit and receive antennas to be collocated at the reader and *bistatic system* with reader transmit and receive antennas to be spaced far apart.

It was observed in Ref. [37] that the average reverse-link interrogation range will be large when monostatic structure is used instead of bistatic structure. However, in rich scattering environments, the bistatic structures are more reliable than the monostatic. In most cases, MIMO system has outperformed its SISO counterpart. For example, 3×3 MIMO-RFID can achieve 60% extra gain in average reverse-link interrogation range compared to that of SISO-RFID system. The reasons could be the exploitation of pinhole diversity and improved SNR received at the reader with multiple antennas. In this work, the main concern was to improve the range rather than identifying multiple tags.

In Ref. [53], Langwieser developed a UHF front end for MIMO applications in RFID. Passive chipped RFID tag has been used for the experiment with one transmitting antenna and two receiving antennas at the reader. Using the measured data, it was claimed that transmit and receive beamforming as well as tag localization can be performed using their front end.

The gains available for chipped RFID tags using multiple antennas are discussed in Ref. [36]. In conventional MIMO, the environment should have rich scattering to exploit spatial multiplexing. In line with that, one can conclude that LOS is not in favor of spatial multiplexing. Even though RFID system channels exhibit LOS propagation, heavy small-scale fading will be present due to indoor operation, a cluttered reader environment, and inhomogeneous nature of the tagged objects [36]. They have studied about an $M \times L \times N$ dyadic backscatter channel, which is a pinhole channel that describes the backscatter propagation radio channel with M transmitter antennas, L RFID tag antennas, and N receiver antennas. This channel was investigated first by Ingram *et al.* [54] with multiple antennas to exploit transmit diversity and spatial multiplexing to increase the range and communication capacity.

In Ref. [36], it was shown that the dyadic backscatter channel has deeper fades than that of the one-way Rayleigh channel but improves as more RFID tag antennas are added. However, pinhole diversity has two advantages over both conventional coherent diversity combining and noncoherent diversity combining. First, it changes the channel distribution to have comparatively less fading. Moreover, diversity gains can be realized in the dyadic backscatter channel using only multiple RFID tag antennas to modulate the backscatter. Hence, no diversity combining is required at the reader, making no changes in the reader receiver hardware, reader transmitter hardware, or signaling scheme. In fact, the actual communication gain in the above dyadic backscatter channel is due to both pinhole diversity gain and increased scattering aperture. In addition, if conventional diversity combining techniques such as maximum ratio combining (MRC) are employed at the reader, even greater gain is achievable. However, it is important to notice that antenna correlation at the tag has to be at a minimum as possible to explore maximum pinhole diversity.

After reviewing the above MIMO-based chipped RFID systems, the following system characteristics were identified. They can be used as design guidelines when developing a MIMO-based chipless RFID system.

- Multiple RF-Tag Antennas. Each antenna of the tag can be used to modulate the interrogating signal.
- Bistatic Reader Architecture. Therefore separate antennas for transmitter and receiver.
- Antenna Arrays Instead of A Signal Antenna. Multiple diversity branches are available.
- Operating Frequency. It should be high enough so that a feasible uncorrelated antenna element spacing at the tag can be achievable.
- Antennas Configuration. Cross-polarized antennas may be used at the reader to reduce self-interference and at the tag to reduce envelop correlation between the signals scattered from each tag antenna. However, using cross-polarized antennas can lead to detrimental effects of unequal diversity branch power.

It is evident that MIMO-based chipped RFID systems have a number of advantages compared to their SISO counterparts. Although they achieve diversity gains from employing multiple antennas, they utilize a microchip in the tag to encode data. The microchip incurs high implementation costs as well as an energy source to power the microchip is needed. Therefore, they do not provide a feasible alternative for optical barcoding that is currently utilized in most mass production item tagging. Chipless RFID systems, on the other hand, do not contain a microchip hence less implementation cost. This is a significant advantage in mass production item tagging. However, there is no reported literature on MIMO-based chipless RFID systems to date except the authors' work in [55]. Thus, there is a need to develop MIMO-based chipless RFID systems that have the potential to replace optical barcode systems used in mass production item tagging.

8.2 REVIEW OF CHIPLESS RFID TAG DETECTION TECHNIQUES

A number of tag detection techniques for chipless RFID systems have been reported. This section summarizes some of the available techniques and compares them in terms of detection accuracy and computation complexity.

The multiresonator-based chipless RFID system presented in Ref. [43] uses a threshold-based tag detection technique, where magnitude of the tag response at resonating frequencies is compared against a threshold level to identify the tag bits. Similar techniques are applied in the chipless RFID systems proposed in Refs [41,44,56–58]. There have been reported tag detection techniques such as Refs [59] and [46], purely based on phase information of the received signal. It was mentioned that information embedded in phase allows to reduce the transmit power compared to magnitude-based detection techniques.

The tag detection performances were able to improve by utilizing a moving average filtering technique in Ref. [42]. The chipless RFID systems proposed in Refs [50,60] utilize both magnitude and phase information for decision making. Utilizing information embedded on both magnitude and phase allows the tag reading to be highly reliable. However, the complete tag detection algorithm was not reported and amalgamating information available in both the magnitude and phase is not known. In addition, a detailed analysis on the performance of the tag detection technique needs to be performed.

The tag detection techniques discussed so far use simple signal processing techniques such as threshold-based detection [43] combined with moving average [42]. The tag detection proposed in Ref. [49] presents an advanced tag detection technique using a continuous wavelet transform. It has managed to overcome the difficulty to detect the signal scattered at the tags with the delay information in the presence of noise. The wavelet transform effectively acts as a matched filter and one benefit of this technique is that a reference tag is not required in advance.

There have been other advanced tag detection techniques such as signal space representation (SSR) reported in Refs [47,61]. The basic principle is to map the received signal vector into a point in an N-dimensional space. All possible tag responses are mapped into fixed points in the N-dimensional space and the minimum distance between the received signal point and other fixed points is calculated to identify the tag bit combination. This is a very accurate technique and it has shown improved performance in successful tag detection. However, one of the challenges is its exponential computation complexity as the number of data bits increases. For example, 20 bit tags have about 1 million unique combinations and the 1 million distances need to be calculated before making a decision.

There have been other techniques reported in Refs [45,48,62–72] using techniques similar to those mentioned earlier. The main limitation of the existing techniques is its low tag reading reliability mostly because of the primitive tag detection techniques used. The advanced tag detection techniques have been able to achieve robust tag reading, however suffer from high computation complexity. Therefore, the motivation is to develop advanced tag detection techniques that can achieve reliable tag reading with relatively low computation complexity. This book presents a maximum likelihood (ML) detection technique with less computation complexity. The following section reviews ML techniques reported in the literature.

8.3 MAXIMUM LIKELIHOOD DETECTION TECHNIQUES

ML detection is a signal processing technique used in communication systems to make decisions by observing a received signal and comparing it with all the possible combinations. An optimal ML detection technique utilizes all aspects of the received signal before making a decision. A suboptimal likelihood detector would use only the main aspects of the received signal such as the magnitude while discarding the phase. Generally, ML-based detection techniques have a very high accuracy. Therefore, ML detection techniques are widely used in communication systems. However, the main drawback is high computation complexity and as a result they are not scalable.

There are a few differences on signals available in communication systems compared to that in a chipless RFID environment. A comparison of the environment available in a communication system and a typical chipless RFID system at 2.4 GHz is shown in Table 8.1.

Signals considered in communication systems are generated based on the modulated bits. Since bits in communication systems are random, the generated signals are also random in nature. On the other hand, tag signals have a limited number of data bits and as a result they can be interrogated multiple times for a higher reading reliability without compromising the reading time. It is similar to a signal obtained by repeating the same bit sequence in a communication system.

Even though the transmitter and receiver are physically separated in communication systems, they are colocated in chipless RFID systems. As a result, the receiver has access to the interrogating signal; hence, a perfect synchronization can be achievable. In communication systems, modulated signals can have intersymbol interference (ISI) due to channel delay spread. However, in frequency domain chipless RFID systems, a bit is modulated using a resonator and its response can interfere with the neighboring resonator responses creating inter-resonator interference (IRI).

Bandwidth in chipless RFID systems are quite high ($> 300\,\mathrm{MHz}$) compared to that of communication systems. As a result, noise level can be expected to be quite high. However, due to short-range operation ($< 1\,\mathrm{m}$) of chipless RFID systems, the received signal power is relatively higher, thanks to the strong LOS presented. Another important observation is the extremely low spectral efficiency of chipless RFID systems as shown in Table 8.1. This is mainly due to the fact that chipless tags have no computation resources and this passive tag designing is deliberately made to be simple. On the other hand, the number of data bits required to transmit is extremely low compared to that of a communication system.

TABLE 8.1   Comparison of Communication System with a Chipless RFID System

Metric	Communication System	Chipless RFID System
Transmitter & receiver	Separated	Colocated
Interference	Intersymbol interference (ISI)	Interresonator interference (IRI)
Bandwidth	Narrowband ($< 5\,\mathrm{MHz}$)	Broadband ($> 300\,\mathrm{MHz}$)
Channel	Multipath propagation	Strong light of sight
Propagation distance	Up to many kilometers	$< 1\,\mathrm{m}$
Spectral efficiency	Several bits per second per Hertz	$\approx 1\,\mathrm{bit}\ 100\,\mathrm{MHz}$ (2.4 GHz)

The understanding of ML detection techniques and the differences in two environments was utilized in proposing likelihood-based detection techniques presented in Chapter 10.

8.4   CONCLUSIONS

A literature survey was carried out in three main categories. The chapter first summarized the available chipless RFID tag types, and multiresonator-based chipless tags were identified as a potential tag type for further investigation. Two tag designs are presented in Chapter 9. Then, available chipped MIMO systems were studied and the main takeaways were discussed. The importance of having multiple antennas both at the tag and the reader was highlighted. Orthogonal antenna configuration is identified as an important aspect when designing a chipless RFID system having multiple antennas. In addition, the operating frequency should be selected such that the uncorrelated antenna element spacing is achievable. These findings will be used in proposing the novel MIMO-based chipless RFID system. Finally, the available tag detection techniques for chipless RFID tag reading were presented and their advantages and limitations were identified. ML-based detection techniques used in communication systems were discussed and difference aspects were compared under a chipless RFID system environment. It was identified that IRI caused is similar to ISI when the resonating frequencies are close to each other in communication systems. Therefore, techniques used in communication systems to mitigate ISI can help in designing tags, which increases data bit capacity in chipless tags. Therefore, the motivation is to remove the guard bands presented between resonance frequencies and mitigate the interference using signal processing techniques at the reader.

In the following chapter, various chipless RFID tag designs and reading methods are presented. Then, the subsequent chapters present the proposed likelihood-based detection techniques of these chipless RFID tags and their tag reading accuracy improvement. At the end of Part II, MIMO signal processing for chipless RFID is presented followed by a chapter focusing on applications of the presented detection techniques.

REFERENCES

1. K. Finkenzeller, *RFID Handbook: Fundamentals and Applications in Contactless Smart Cards and Identification*, 2nd ed. New York, NY, USA: John Wiley & Sons, Inc., 2003.
2. S. Preradovic and N. Karmakar, *Modern RFID readers*, 2007 (*accessed December 24, 2014*). [Online]. Available: http://www.microwavejournal.com/articles/5271-modern-rfid-readers.
3. S. Preradovic and N. Karmakar, "Chipless RFID: bar code of the future", *IEEE Microwave Magazine*, vol. 11, no. 7, pp. 87–97, Dec 2010.
4. U. Kaiser and W. Steinhagen, "A low power transponder IC for high performance identification systems," in *Custom Integrated Circuits Conference, 1994., Proceedings of the IEEE 1994*, San Diego, CA, USA, May 1994, pp. 335–338.

5. S. Preradovic, S. Roy, and N. Karmakar, "RFID system based on fully printable chipless tag for Paper-/Plastic-Item tagging," *IEEE Antennas and Propagation Magazine*, vol. 53, no. 5, pp. 15–32, Oct 2011.

6. F. Kamoun, "RFID system management: state-of-the art and open research issues," *IEEE Transactions on Network and Service Management*, vol. 6, no. 3, pp. 190–205, Sept 2009.

7. X. Chen, W. G. Yeoh, Y. B. Choi, H. Li, and R. Singh, "A 2.45-GHz near-field RFID system with passive On-Chip antenna tags," *IEEE Transactions on Microwave Theory and Techniques*, vol. 56, no. 6, pp. 1397–1404, June 2008.

8. A. Shameli, A. Safarian, A. Rofougaran, M. Rofougaran, J. Castaneda, and F. De Flaviis, "A UHF near-field RFID system with fully integrated transponder," *IEEE Transactions on Microwave Theory and Techniques*, vol. 56, no. 5, pp. 1267–1277, May 2008.

9. A. Athalye, V. Savic, M. Bolic, and P. Djuric, "Novel semi-passive RFID system for indoor localization," *IEEE Sensors Journal*, vol. 13, no. 2, pp. 528–537, Feb 2013.

10. Y. Weng, S. Cheung, T. Yuk, and L. Liu, "Design of chipless UWB RFID system using A CPW multi-resonator," *IEEE Antennas and Propagation Magazine*, vol. 55, no. 1, pp. 13–31, Feb 2013.

11. C. Floerkemeier and S. Sarma, "An overview of RFID system interfaces and reader protocols," in *RFID, 2008 IEEE International Conference on*, April 2008, pp. 232–240.

12. H. Song, Y. Zhao, and H. Lan, "Multi-channel active RFID system for IoT application," in *Measurement, Information and Control (MIC), 2012 International Conference on*, vol. 2, May 2012, pp. 568–571.

13. G. Fritz, V. Beroulle, M. Nguyen, O. Aktouf, and I. Parissis, "Read-error-rate evaluation for RFID system on-line testing," in *Mixed-Signals, Sensors and Systems Test Workshop (IMS3TW), 2010 IEEE 16th International*, June 2010, pp. 1–6.

14. S. Sabesan, M. Crisp, R. Penty, and I. White, "An error free passive UHF RFID system using a new form of wireless signal distribution," in *RFID (RFID), 2012 IEEE International Conference on*, April 2012, pp. 58–65.

15. S. Ma, Y. Zhang, and D. Wang, "Distributed work flow system for RFID integration and application," in *Information, Computing and Telecommunication, 2009. YC-ICT '09. IEEE Youth Conference on*, Sept 2009, pp. 62–65.

16. S. Mirshahi, S. Uysal, and A. Akbari, "Integration of RFID and WSN for supply chain intelligence system," in *Electronics, Computers and Artificial Intelligence (ECAI), 2013 International Conference on*, June 2013, pp. 1–6.

17. R. Falk, F. Kohlmayer, A. Koepf, M. Braun, H. Seuschek, and M. Li, "Application of passive asymmetric RFID tags in a high-assurance avionics multi-domain RFID processing system," in *RFID Systems and Technologies (RFID SysTech), 2008 4th European Workshop on*, June 2008, pp. 1–7.

18. S.-K. Youm, J.-H. Kim, and S.-K. Cho, "A study on the methodology for testing of RFID system at library," in *Multimedia and Ubiquitous Engineering, 2007. MUE '07. International Conference on*, April 2007, pp. 1076–1079.

19. I. Mun, A. Kantrowitz, P. Carmel, K. Mason, and D. Engels, "Active RFID system augmented with 2D barcode for asset management in a hospital setting," in *RFID, 2007. IEEE International Conference on*, March 2007, pp. 205–211.

20. A. Fawky, M. Mohammed, M. El-Hadidy, and T. Kaiser, "UWB chipless RFID system performance based on real world 3D-deterministic channel model and ZF equalization," in *Antennas and Propagation (EuCAP), 2014 8th European Conference on*, April 2014, pp. 1765–1768.

21. Z. J. Guo, "Research on test technology of RFID system," in *Control and Decision Conference (CCDC), 2013 25th Chinese*, May 2013, pp. 2793–2796.

22. K. Kozlowski, L. Kulas, and K. Nyka, "New RFID readers for scalable RFID system," in *Information Technology (ICIT), 2010 2nd International Conference on*, June 2010, pp. 95–98.

23. N. Aziz, I. Alias, A. Hashim, R. Mustafa, K. Anuar, S. Ahmad, and W. Muhamad, "Smart RFID system for oil palm bio-laboratory," in *RF and Microwave Conference, 2008. RFM 2008. IEEE International*, Dec 2008, pp. 247–251.

24. S.-Y. Chan, S.-W. Luan, J.-H. Teng, and M.-C. Tsai, "Design and implementation of a RFID-based power meter and outage recording system," in *Sustainable Energy Technologies, 2008. ICSET 2008. IEEE International Conference on*, Nov 2008, pp. 750–754.

25. C.-T. Huang, S.-J. Wang, W.-L. Wang, and Y.-S. Wang, "Construction of an online RFID enabled supply chain system reliability monitoring model," in *Computer, Consumer and Control (IS3C), 2012 International Symposium on*, June 2012, pp. 626–629.

26. M. Nassih, I. Cherradi, Y. Maghous, B. Ouriaghli, and Y. Salih-Alj, "Obstacles recognition system for the blind people using RFID," in *Next Generation Mobile Applications, Services and Technologies (NGMAST), 2012 6th International Conference on*, Sept 2012, pp. 60–63.

27. J.-S. Cho, S.-C. Kim, and S.-S. Yeo, "RFID system security analysis, response strategies and research directions," in *Parallel and Distributed Processing with Applications Workshops (ISPAW), 2011 Ninth IEEE International Symposium on*, May 2011, pp. 371–376.

28. Y. Zuo, "Survivable RFID systems: issues, challenges, and techniques," *IEEE Transactions on Systems, Man, and Cybernetics, Part C: Applications and Reviews*, vol. 40, no. 4, pp. 406–418, July 2010.

29. E. Hagras, "Interleave division multiple access; robust anti-collision protocol for UWB RFID system in non-Gaussian impulsive channel," in *Radio Science Conference (NRSC), 2014 31st National*, April 2014, pp. 202–209.

30. Z. Al-Amir, F. Al-Saidi, and H. Abdulkadir, "Design and implementation of RFID system," in *Systems, Signals and Devices, 2008. IEEE SSD 2008. 5th International Multi-Conference on*, July 2008, pp. 1–6.

31. M. Mubarak, J. Manan, and S. Yahya, "Trusted anonymizer-based RFID system with integrity verification," in *Information Assurance and Security (IAS), 2011 7th International Conference on*, Dec 2011, pp. 98–103.

32. A. Bhattacharjya and R. Pal, "Distributed design of universal lightweight RFID system for large-scale RFID operation," in *Business Innovation and Technology Management (APBITM), 2011 IEEE International Summer Conference of Asia Pacific*, July 2011, pp. 40–44.

33. S. Sabesan, M. Crisp, R. Penty, and I. White, "Wide area passive UHF RFID system using antenna diversity combined with phase and frequency hopping," *IEEE Transactions on Antennas and Propagation*, vol. 62, no. 2, pp. 878–888, Feb 2014.

34. D.-G. Min, J.-W. Kim, and M.-S. Jun, "The entrance authentication system in real-time using face extraction and the RFID tag," in *Ubiquitous Computing and Multimedia Applications (UCMA), 2011 International Conference on*, April 2011, pp. 20–24.

35. H. Cheng, W. Ni, and N. Li, "A systematic scheme for designing RFID systems with high object detection reliability," in *Information Science, Electronics and Electrical Engineering (ISEEE), 2014 International Conference on*, vol. 3, April 2014, pp. 1521–1526.

36. J. Griffin and G. Durgin, "Gains for RF tags using multiple antennas," *IEEE Transactions on Antennas and Propagation*, vol. 56, no. 2, pp. 563–570, Feb 2008.

37. D.-Y. Kim, H.-S. Jo, H. Yoon, C. Mun, B.-J. Jang, and J.-G. Yook, "Reverse-link interrogation range of a UHF MIMO-RFID system in Nakagami- m fading channels," *IEEE Transactions on Industrial Electronics*, vol. 57, no. 4, pp. 1468–1477, April 2010.

38. N. Saldanha and D. Malocha, "Low loss SAW RF ID tags for space applications," in *Ultrasonics Symposium, 2008. IUS 2008. IEEE*, Beijing, China, Nov 2008, pp. 292–295.

39. N. I. of Advanced Industrial Science and T. (AIST), Printing of organic thin-film transistor arrays on flexible substrates, 2008 (accessed December 24, 2014). [Online]. Available: http://www.idtechex.com/research/reports/rfid_forecasts_players_and_opportunities_2006 _2016_000137.asp.

40. J. Vemagiri, A. Chamarti, M. Agarwal, and K. Varahramyan, "Transmission line delay-based radio frequency identification (RFID) tag," *Microwave and Optical Technology Letters*, vol. 49, no. 8, pp. 1900–1904, 2007.

41. A. Vena, E. Moradi, K. Koski, A. Babar, L. Sydanheimo, L. Ukkonen, and M. Tentzeris, "Design and realization of stretchable sewn chipless RFID tags and sensors for wearable applications," in *RFID (RFID), 2013 IEEE International Conference on*, April 2013, pp. 176–183.

42. R. Koswatta and N. Karmakar, "Moving average filtering technique for signal processing in digital section of UWB chipless RFID reader," in *Microwave Conference Proceedings (APMC), 2010 Asia-Pacific*, Dec 2010, pp. 1304–1307.

43. S. Preradovic, I. Balbin, N. Karmakar, and G. Swiegers, "Multiresonator-based chipless RFID system for low-cost item tracking," *IEEE Transactions on Microwave Theory and Techniques*, vol. 57, no. 5, pp. 1411–1419, May 2009.

44. M. Bhuiyan and N. Karmakar, "Chipless RFID tag based on split-wheel resonators," in *Antennas and Propagation (EuCAP), 2013 7th European Conference on*, April 2013, pp. 3054–3057.

45. A. Blischak and M. Manteghi, "Pole residue techniques for chipless RFID detection," in *Antennas and Propagation Society International Symposium, 2009. APSURSI '09. IEEE*, Charleston, SC, USA, June 2009, pp. 1–4.

46. S. Mukherjee, "Chipless radio frequency identification by remote measurement of complex impedance," in *Microwave Conference, 2007. European*, Munich, Germany, Oct 2007, pp. 1007–1010.

47. P. Kalansuriya, N. Karmakar, and E. Viterbo, "On the detection of frequency-spectra-based chipless RFID using UWB impulsed interrogation," *IEEE Transactions on Microwave Theory and Techniques*, vol. 60, no. 12, pp. 4187–4197, Dec 2012.

48. M. Manteghi, "A novel approach to improve noise reduction in the Matrix Pencil Algorithm for chipless RFID tag detection," in *Antennas and Propagation Society International Symposium (APSURSI), 2010 IEEE*, Toronto, ON, Canada, July 2010, pp. 1–4.

49. A. Lazaro, A. Ramos, D. Girbau, and R. Villarino, "Chipless UWB RFID tag detection using continuous wavelet transform," *IEEE Antennas and Wireless Propagation Letters*, vol. 10, pp. 520–523, 2011.

50. A. Vena, E. Perret, and S. Tedjini, "Chipless RFID tag using hybrid coding technique," *IEEE Transactions on Microwave Theory and Techniques*, vol. 59, no. 12, pp. 3356–3364, Dec 2011.

51. D. Chizhik, G. Foschini, and R. Valenzuela, "Capacities of multi-element transmit and receive antennas: correlations and keyholes," *Electronics Letters*, vol. 36, no. 13, pp. 1099–1100, Jun 2000.

52. M. Nakagami, "The m-distribution -A general formula of intensity distribution of rapid fading," in *Statistical Methods in Radio Wave Propagation*, W. Hoffman, Ed. Pergamon, 1960, pp. 3–36.

53. R. Langwieser, C. Angerer, and A. Scholtz, "A UHF frontend for MIMO applications in RFID," in *Radio and Wireless Symposium (RWS), 2010 IEEE*, New Orleans, VA, USA, Jan 2010, pp. 124–127.

54. M. A. Ingram, M. F. Demirkol, and D. Kim, "Transmit diversity and spatial multiplexing for RF links using modulated backscatter," in *International Symposium on Signals, Systems and Electronics (ISSSE), 2001*, Tokyo, Japan, July 2001.

55. C. Divarathne and N. Karmakar, "MIMO based chipless RFID system," in *RFID-Technologies and Applications (RFID-TA), 2012 IEEE International Conference on*, Nov 2012, pp. 423–428.

56. M. Islam and N. Karmakar, "A novel compact printable dual-polarized chipless RFID system," *IEEE Transactions on Microwave Theory and Techniques*, vol. 60, no. 7, pp. 2142–2151, July 2012.

57. A. Vena, E. Perret, and S. Tedjni, "A depolarizing chipless RFID tag for robust detection and its FCC compliant UWB reading system," *IEEE Transactions on Microwave Theory and Techniques*, vol. 61, no. 8, pp. 2982–2994, Aug 2013.

58. A. Vena, E. Perret, and S. Tedjini, "A fully printable chipless RFID tag with detuning correction technique," *IEEE Microwave and Wireless Components Letters*, vol. 22, no. 4, pp. 209–211, April 2012.

59. I. Balbin and N. Karmakar, "Phase-encoded chipless RFID transponder for large-scale low-cost applications," *IEEE Microwave and Wireless Components Letters*, vol. 19, no. 8, pp. 509–511, Aug 2009.

60. R. Koswatta and N. Karmakar, "A novel reader architecture based on UWB chirp signal interrogation for multiresonator-based chipless RFID tag reading," *IEEE Transactions on Microwave Theory and Techniques*, vol. 60, no. 9, pp. 2925–2933, Sept 2012.

61. R. Rezaiesarlak and M. Manteghi, "A space-frequency technique for chipless RFID tag localization," *IEEE Transactions on Antennas and Propagation*, vol. 62, no. 11, pp. 5790–5797, Nov 2014.

62. R. Rezaiesarlak and M. Manteghi, "A new detection technique for identifying chipless RFID tags," in *Radio Science Meeting (USNC-URSI NRSM), 2014 United States National Committee of URSI National*, Boulder, CO, USA, Jan 2014, pp. 1.

63. R. Rezaiesarlak and M. Manteghi, "Time-frequency analysis of the scattered signal from chipless RFID tags," in *Radio Science Meeting (USNC-URSI NRSM), 2014 United States National Committee of URSI National*, Boulder, CO, USA, Jan 2014, pp. 1.

64. P. Kalansuriya, N. Karmakar, and E. Viterbo, "Signal space representation of chipless RFID tag frequency signatures," in *Global Telecommunications Conference (GLOBE-COM 2011), 2011 IEEE*, Houston, TX, USA, Dec 2011, pp. 1–5.

65. R. Koswatta and N. Karmakar, "A novel method of reading multi-resonator based chipless RFID tags using an UWB chirp signal," in *Microwave Conference Proceedings (APMC), 2011 Asia-Pacific*, Melbourne, Australia, Dec 2011, pp. 1506–1509.

66. A. Blischak and M. Manteghi, "Embedded singularity chipless RFID tags," *IEEE Transactions on Antennas and Propagation*, vol. 59, no. 11, pp. 3961–3968, Nov 2011.

67. S. Mukherjee, "Chipless radio frequency identification by remote measurement of complex impedance," in *Wireless Technologies, 2007 European Conference on*, Munich, Germany, Oct 2007, pp. 249–252.

68. W. Dullaert, L. Reichardt, and H. Rogier, "Improved detection scheme for chipless RFIDs using prolate spheroidal wave function-based noise filtering," *IEEE Antennas and Wireless Propagation Letters*, vol. 10, pp. 472–475, 2011.

69. P. Kalansuriya and N. Karmakar, "Time domain analysis of a backscattering frequency signature based chipless RFID tag," in *Microwave Conference Proceedings (APMC), 2011 Asia-Pacific*, Melbourne, Australia, Dec 2011, pp. 183–186.

70. M. Manteghi, "A space-time-frequency target identification technique for chipless RFID applications," in *Antennas and Propagation (APSURSI), 2011 IEEE International Symposium on*, Spokane, WA, USA, July 2011, pp. 3350–3351.

71. F. Costa, S. Genovesi, and A. Monorchio, "A chipless RFID based on multiresonant high-impedance surfaces," *IEEE Transactions on Microwave Theory and Techniques*, vol. 61, no. 1, pp. 146–153, Jan 2013.

72. V. Montilla, E. Ramon, and J. Carrabina, "Frequency scan technique for inkjet-printed chipless sensor tag reading," in *Electronics, Circuits, and Systems (ICECS), 2010 17th IEEE International Conference on*, Athens, Greece, Dec 2010, pp. 1100–1103.

9

CHIPLESS RFID TAG DESIGN

9.1 INTRODUCTION

In this chapter, the physical layer developments of the two chipless radiofrequency identification (RFID) tag types, experimental setup, and results are presented in microwave and millimeter-wave domains. The first tag type is a single input single output (SISO) tag having circular resonators that operates between 21 and 27 GHz. The second type is the novel multiple input multiple output (MIMO) tag having multiresonators that operates at the frequency band 2.4 GHz. A brief summary of the two tag types are given in Figure 9.1. The rest of the chapter describes the design of the tags and experimental verification of the proposed tag detection techniques in *Chapter 10* and *11*.

9.2 SISO TAG DESIGN

In this section, a circular patch resonator-based backscattering chipless RFID tag design is discussed. The tags are operating between 21 and 27 GHz frequency range. The tags are fabricated on a thin film paper using the SATO printer having a conductive ink. Then, the tags are read using an existing chipless RFID reader developed by Kalansuriya at Monash Microwave, Antenna, RFID and Sensor Laboratory, Monash University.

Advanced Chipless RFID: MIMO-Based Imaging at 60 GHz – ML Detection, First Edition.
Nemai Chandra Karmakar, Mohammad Zomorrodi, and Chamath Divarathne.
© 2016 John Wiley & Sons, Inc. Published 2016 by John Wiley & Sons, Inc.

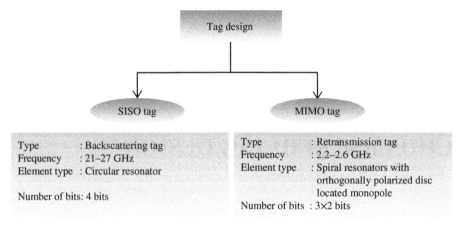

Figure 9.1 Tag types used in SISO and MIMO detection algorithm developments.

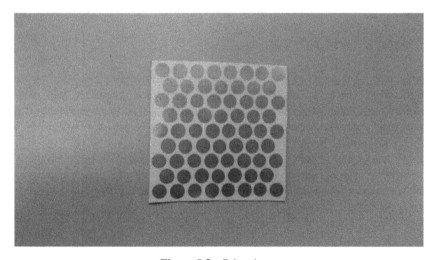

Figure 9.2 Printed tag.

9.2.1 Tag Design and Fabrication

Figure 9.2 shows the photograph of a 4-bit tag designed at the 21–27 GHz ultrawide band (UWB). The tag is comprised of four sets of distinct resonators thoughtfully located so that nearby resonances are not interfered. The tag is designed to occupy 4 bits in the 21–27 frequency band. The resonance frequencies are selected as 22.5, 23.5, 24.5, and 25.5 GHz as simulation shows that the 3-dB bandwidth of each resonator response at this frequency band is about 1 GHz. The diameter of the circle is inversely proportional to the resonance frequency. The diameters for the above resonance frequencies are found and displayed in Table 9.1.

TABLE 9.1 Simulation Parameters

Resonance Frequency (GHz)	Diameter of the Circular Patch (mm)
22.5	4.38
23.5	4.59
24.5	4.84
25.5	5.13

The specific frequency resonators are repeated to increase the backscattered signals of the individual resonators [1]. The tag is printed in-home using MMARS Laboratory's SATO printer. This monochrome printer has a resolution of 600 dpi. The substrate is 0.09-mm-thick glossy paper with a bulk silver coating of 10 μm on top. The design is thermally transferred to the paper substrate and the fully printable silver tag is created on the paper substrate.

9.2.2 Experimental Setup

In this section, the experimental setup used to verify the performance of the tags was designed. A chipless RFID reader that reads the magnitude of the tag response was used to read the tag designed in the previous section. An experiment was set up as shown in Figure 9.3. The reader transmits a narrow banded sinusoid signal using a transmit horn antenna and as shown in the figure the receiver receives the tag response for that frequency using a receive horn antenna. The magnitude of the received signal is recorded along with the frequency of the signal sent. Similarly, the frequency is swept across the 21–27 GHz band and the complete tag response is recorded.

Figure 9.4 shows the tag response ([1111]) recorded by the chipless RFID reader. It can be seen that the four resonances occur at the designed resonating frequencies. The presence (ON) and absence (OFF) of the resonators can be selected to form 16 different tag types for a 4-bit tag as shown in Table 9.2 The circular patches having the same diameter increase the effective radar cross section and hence form a bigger frequency dip at the resonance frequency. The four resonator sets are selected such that tags with all 16 combinations can be designed. These 16 tag types are fabricated and used for testing the detection techniques derived in *Chapter 10*.

9.3 MIMO TAG DESIGN

One of the original contributions of the proposed MIMO-based chipless RFID system is the MIMO tag design. The main parts of the proposed chipless MIMO tag are identified as the power divider, monopole tag antennas, and multispiral resonators. These individual components are designed in computer simulation tool (CST) and their performance is verified. Following are the detailed descriptions of the component-level design and their integration to a complete MIMO tag.

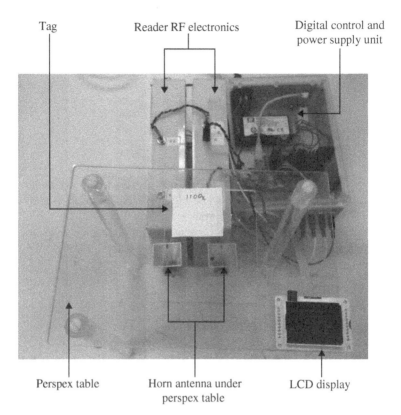

Figure 9.3 Experimental setup.

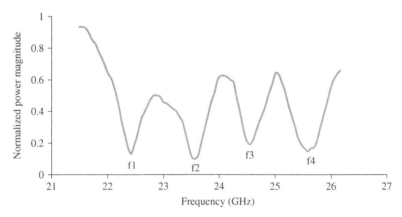

Figure 9.4 Magnitude of the tag response for tag [1111].

TABLE 9.2   Tag Types Given by Resonator Combinations

Tag Type	Resonator Combinations			
	f1	f2	f3	f4
[0000]	OFF	OFF	OFF	OFF
[0001]	OFF	OFF	OFF	ON
[0010]	OFF	OFF	ON	OFF
[0011]	OFF	OFF	ON	ON
[0100]	OFF	ON	OFF	OFF
[0101]	OFF	ON	OFF	ON
[0110]	OFF	ON	ON	OFF
[0111]	OFF	ON	ON	ON
[1000]	ON	OFF	OFF	OFF
[1001]	ON	OFF	OFF	ON
[1010]	ON	OFF	ON	OFF
[1011]	ON	OFF	ON	ON
[1100]	ON	ON	OFF	OFF
[1101]	ON	ON	OFF	ON
[1110]	ON	ON	ON	OFF
[1111]	ON	ON	ON	ON

9.3.1   Power Divider Design

The power divider needs to divide the power equally into two branches over a broadband of 400 MHz centered at 2.4 GHz. Therefore, a symmetrical T-junction power divider is designed. The design parameters are the length and the width of each transmission line presented in the divider. The designed and fabricated T-junction power divider is shown in Figure 9.5. The power divider is designed on TLX8 substrate with a relative permittivity (ϵ_r) of 2.4, loss tangent ($\tan \delta$) of 0.004, and thickness (h) of 0.5 mm. Both the simulated and the experimental S parameter magnitudes versus frequency for the power divider are shown in Figure 9.6. It is clear that both simulated and measured S21 and S31 are similar and close to the required value of -3 dB (half power division). The measured and simulated S11 in decibels (return loss) versus frequency is also shown in the figure. There is a difference in the two curves and this is possibly due to fabrication defects, where it shifts the operating frequency little higher. As far as the performance is concerned, the fabricated power divider still has a bandwidth of over 500 MHz at -15 dB centered around 2.4 GHz. So the power divider performance is acceptable and any further tuning is not attempted. Once the satisfactory performance of the T-junction power divider is obtained, the next component designed is the monopole antenna.

9.3.2   Monopole Antenna Design

A monopole antenna is designed to operate at 2.4 GHz with a bandwidth of 400 MHz. The monopole antenna is selected due to its figure-of-eight bream radiation pattern and distinct polarization so that vertical and horizontal antennas can be used to reduce

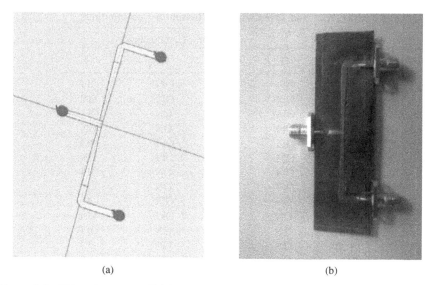

(a)   (b)

Figure 9.5 T-junction power divider. (a) CST design; (b) fabricated power divider. *TLX8 substrate with $\epsilon_r = 2.4$, $\tan \delta = 0.004$ and $h = 0.5\,mm$.*

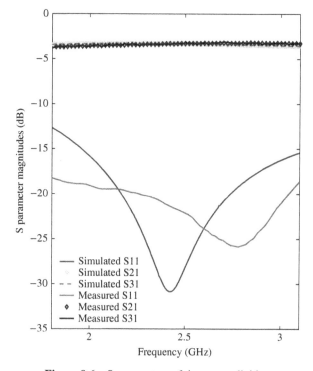

Figure 9.6 S-parameters of the power divider.

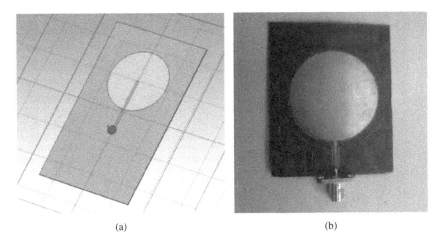

(a)   (b)

Figure 9.7 Monopole antenna. (a) CST design; (b) fabricated monopole).

the cross talk between the transmit and receive chains of the MIMO chipless RFID tag. A microstrip-based antenna with a circular patch is used for size optimization [2]. Figure 9.7 shows the monopole antenna design in CST as well as the fabricated antenna. The design parameters are the radius of the circular patch, gap between the ground plane, and disc edge and the dimension of the feedline [3].

The performance of the monopole antennas is presented next. Figure 9.8 shows the return loss versus frequency of the antenna obtained from both simulated and

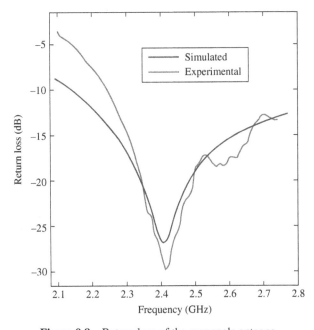

Figure 9.8 Return loss of the monopole antenna.

Farfield realized gain Abs (ϕ = 90)

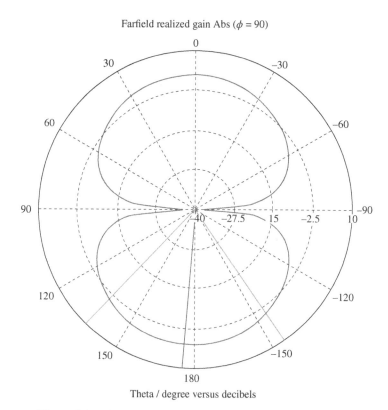

Theta / degree versus decibels

Figure 9.9   Simulated radiation pattern of the monopole antenna.

measured data. It can be seen that they both agree and the antenna has an operating bandwidth of over 400 MHz at −10 dB.

Figure 9.9 shows the antenna radiation pattern at 2.4 GHz, which is a figure-of-eight omnidirectional radiation pattern with the maximum gain of the main lobe with 2.7 dB, which is acceptable.

The realized gain of the antenna on the main lobe was analyzed over a bandwidth of 400 MHz centered around 2.4 GHz and the results are shown in Figure 9.10. It is clear that the realized gain is above 2.5 dB over the frequency band. These results verify the successful operation of the monopole antenna at a bandwidth of 400 MHz centered around 2.4 GHz. After obtaining satisfactory performance from the designed antenna, the spiral resonators are designed.

9.3.3   Spiral Resonator Design

A set of spiral resonators are designed to operate at the resonance frequencies given by Table 9.1. Design parameters are the transmission line lengths, widths, and the gap between the microstrip lines [3]. Figure 9.11 shows both the CST designs on spiral resonators and one of the fabricated resonators.

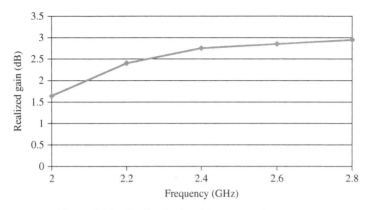

Figure 9.10 Realized gain of the monopole antenna.

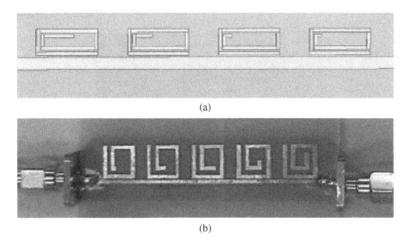

(a)

(b)

Figure 9.11 Spiral resonators. (a) CST design; (b) fabrication.

A spiral resonator response obtained from CST simulations is shown in Figure 9.12. It is clearly seen that both the amplitude and the phase contain the data and existing techniques mostly rely on only one of them. However, the proposed detection techniques use the information available in both amplitude and phase. A photograph of the fabricated MIMO tag with all "0" bits (no multiresonator) in branch connected to Tx_1 and all "1" bits in the branch connected to Tx_2 is shown in Figure 9.13.

9.3.4 Experimental Setup

The schematic of the experiment conducted to measure the MIMO tag response using an arbitrary waveform generator (AWG) and an oscilloscope with a high sampling

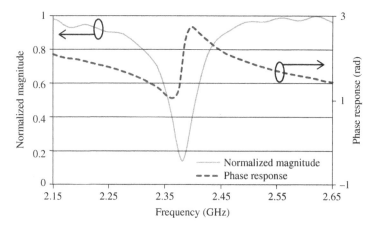

Figure 9.12   CST-generated resonator response.

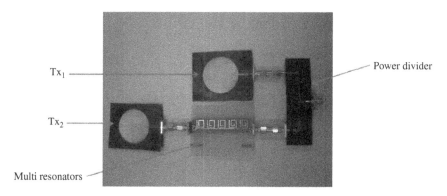

Figure 9.13   Fabricated MIMO tag.

rate is shown in Figure 9.14. However, antennas are replaced using cables as the sole purpose of this work is to verify the validity of the ML-based detection method.

Figure 9.15 shows the CST-generated tag response and the measured tag response for tag bits [1010]. S_{21} measurements across the resonator are recorded using a performance network analyzer and converted from log scale to linear scale for comparison with the CST-generated response. It can be seen that they are closely matched.

So far the design of a MIMO-based chipless RFID tag is discussed. Its individual component performances are analyzed using the CST-simulated results and experimental results. It is clear from Figure 9.15 that there are some anomalies between the theoretical and experimental results. This may be because of the coupling between the two branches in the MIMO tag. However, the proposed MIMO-based chipless RFID system has an inbuilt mechanism to overcome this challenge. Any coupling

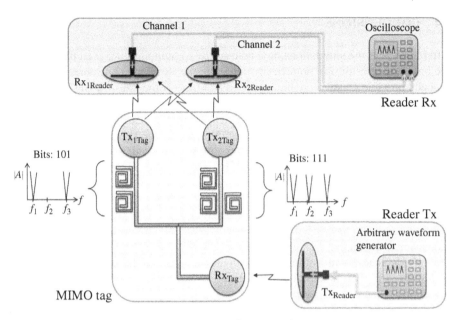

Figure 9.14 MIMO tag experiment.

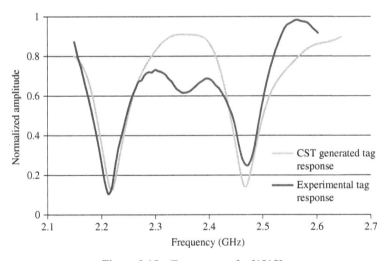

Figure 9.15 Tag response for [1010].

between the antennas can be incorporated into the 2×2 MIMO channel matrix. This can be performed using a calibration tag. Now the system is free from any coupling between the two branches, and with the known channel, it is possible to perform tag detection.

9.4 CONCLUSIONS

After analyzing the simulations, it is noteworthy to pinpoint that, even though there are only two transmitting branches presented in the RFID tag considered, it is theoretically possible to add more branches and still recover the transmitted signals given that the number of receiving antennas in the reader is larger than or equal to the number of transmitting branches in the tag. Hence, without increasing the bandwidth, the bit capacity can be further increased using the same frequency resonators compared with having only one branch at the tag. However, it is required to evaluate the effect of mutual coupling between the antennas with higher number of transmitting branches in the tag.

In the RFID tag proposed, there is only one receiving antenna through which the received signal will be divided into two equal components. The proposed concept can be extended to having a dedicated receiving antenna for each component, hence increasing the effective signal-to-noise ratio (SNR) at each branch. Therefore, with multiple dedicated transmitting and receiving antennas on the tag performance can be improved further. In addition, the concept can be further extended to multiple tag detection if each branch is considered as a separate tag.

Furthermore, the use of IQ modulation/demodulation allows an extra degree of freedom to increase the bit capacity. Since the baseband signal considered is complex, it is possible to have asymmetric frequency response in positive and negative frequencies. Therefore, the eligible frequency band in the passband centered around the radiofrequency (RF) carrier doubles, allowing more resonators to be placed in the tag, without increasing the sampling rate of the ADC at the receiving end of the reader. After analyzing the above results, it can be concluded that MIMO is a competitive candidate for improving reliability or the bit capacity of a resonator-based chipless RFID system.

REFERENCES

1. P. Kalansuriya, N. Karmakar, and E. Viterbo, "On the detection of frequency-spectra-based chipless RFID using UWB impulsed interrogation," *IEEE Transactions on Microwave Theory and Techniques*, vol. 60, no. 12, pp. 4187–4197, 2012.

2. S. Preradovic, I. Balbin, N. Karmakar, and G. Swiegers, "Multiresonator-based chipless RFID system for low-cost item tracking," *IEEE Transactions on Microwave Theory and Techniques*, vol. 57, no. 5, pp. 1411–1419, May 2009.

3. N. Karmakar, S. Roy, S. Preradovic, T. Vo, and S. Jenvey, "Development of low-cost active RFID tag at 2.4 GHz," in Microwave Conference, 2006. 36th European, 2006, pp. 1602–1605.

10

ML DETECTION TECHNIQUES FOR SISO CHIPLESS RFID TAGS

10.1   INTRODUCTION

Research on chipless radio frequency identification (RFID) systems are mainly emphasizing on improving the RFID reader architecture and the chipless tag design. As a result, tag detection techniques are overshadowed, hence they are using primitive signal processing techniques. The main focus of this chapter is to improve the signal processing techniques using the same reader architecture and tag design. Therefore, the proposed tag detection techniques are compatible with the existing RFID systems. The improvements are expected in successful tag detection rate and the tag reading range.

The existing chipless signal processing techniques for tag detection is as primitive as threshold-based detection. Maximum likelihood (ML)-based detection techniques have shown improved performances in communication systems over primitive techniques such as threshold-based detection techniques. The motivation for this work is to apply the ML detection techniques for chipless RFID tag detection so that the existing RFID systems would produce better results in terms of the detection error rate (DER) and the reading range.

The rest of the chapter is organized as follows. First, the theory behind deriving four ML expressions for a single-input single-output (SISO)-based chipless RFID system is presented. The different expressions are derived based on the availability of the channel information (known or unknown channel) and real or complex signal processing. Then, a computationally feasible tag detection technique is presented so

Advanced Chipless RFID: MIMO-Based Imaging at 60 GHz–ML Detection, First Edition.
Nemai Chandra Karmakar, Mohammad Zomorrodi, and Chamath Divarathne.
© 2016 John Wiley & Sons, Inc. Published 2016 by John Wiley & Sons, Inc.

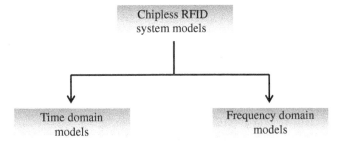

Figure 10.1   RFID system models.

that the detection technique can be implemented on a portable RFID reader. Next, the detection techniques are implemented in MATLAB and its results are compared. Finally, the original contribution and a discussion on each detection technique are presented in the conclusion section.

The system models presented in the chapter are for frequency-domain-based chipless RFID tags. However, the models are based on either signal using time-domain samples or frequency-domain samples. Hence, the system models presented can be categorized into two sections as shown in Figure 10.1.

10.2   SYSTEM MODELS–TIME DOMAIN

A multiresonator-based chipless RFID system consists of three main components, namely, a reader, a tag, and the middleware as shown in Figure 10.2. The RFID reader generates an interrogating signal and transmits toward the tag using a transmitting antenna (Tx_r). The interrogating signal will be received by the tag using its dedicated receiving antenna (Rx_t), which has the same polarization as the transmitting antenna

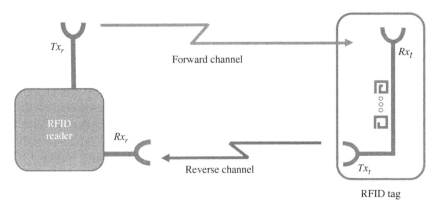

Figure 10.2   Overview of chipless RFID system.

of the reader. Then the received signal propagates via a frequency modulation circuit that comprises a cascade of spiral resonators. Depending on the resonator combinations (presence and absence of resonators), a unique tag response is available at the end of the microstrip line. This process is called *tag modulation* from here onward. Then, the tag response will be transmitted using the dedicated transmitting antenna of the tag (Tx_t). The polarization of the transmitting antennas is orthogonal to that of the receiving antenna of the tag. The transmitted tag response is received at the reader using antennas having matching polarity. This careful selection of the antenna configuration limits any unwanted cross-coupling between antennas.

Next, the system is modeled using a number of system models in the following sections.

10.2.1 System Model I–Real Signals

The system described in Figure 10.2 is modeled first using the following simple signal model. Later, more assumptions are relieved and a comprehensive analysis is performed so that the new models closely describe the real system.

The signals considered in *System Model I* are assumed to be real signals, meaning the received signal is directly sampled at a very high rate. First, the interrogating signal transmitted from the RFID reader reaches the tag through the forward channel as shown in Figure 10.2. Then the signal received by the tag is modulated by the resonator combination present in the tag. $\bar{S}_m$ is defined as this resultant signal available for transmitting back toward the reader. $\bar{S}_m$ is a unique tag response for a given tag m and it is a vector having a length of N. $\bar{S}_m$ is then transmitted from the tag to the RFID reader via the reverse channel as shown in Figure 10.2.

Both the forward and reverse channels are in short range with a strong line of sight. The channels are assumed to be real and known constants. Mixing with the forward channel, tag modulation and mixing with the reverse channel happen in a cascaded manner. As a result, the product of both channels can be represented using a real constant h. The received signal at the reader is added with noise ($\bar{\omega}$) produced by the receiver circuit at the RFID reader. The resultant signal is called $\bar{y}$ and can be represented using Equation (10.1)

$$\bar{y} = h\bar{S}_m + \bar{\omega} \tag{10.1}$$

When the RFID reader transmits the interrogating signal, it is first received by the receiving antenna of the tag. Then the signal is modulated by the tag and transmitted back toward the RFID reader. $\bar{S}_m$ includes both the resonator response as well as the noise introduced by the tag antennas. Therefore, $\bar{S}_m$ is actually dependent on the tag combination. However, the amplitude of the noise added by the receiving antenna of the RFID reader is much higher than the noise added by the tag antennas. This happens due to the close proximity of the transmitter and receiver electronics of the reader to the reader antenna. Also, the reader antenna is expected to have a higher gain compared to the tag antennas. Therefore, it may also pick more surrounding noises. Hence, noise presented in $\bar{S}_m$ can be neglected and $\bar{S}_m$ can be treated as the pure tag response. The noise presented at the received signal $\bar{y}$ is only due to the noise added at the reader. Therefore, noise $\bar{\omega}$ added at the reader can be assumed to be independent of the tag response $\bar{S}_m$. In addition, individual time samples of $\bar{\omega}$ vector is assumed to follow an independent and identical Gaussian distribution (i.i.d.) with zero mean and a variance of σ_ω^2. As a result, the distributions of $\bar{\omega}$ and $\bar{y}$ can be derived as in Equation (10.2).

$$\bar{\omega} \sim N\ (\bar{0}, \sigma_\omega^2 I_N)$$
$$\bar{y} \sim N\ (h\bar{S}_m, \sigma_\omega^2 I_N) \tag{10.2}$$

where I_N is the identity matrix with $N \times N$ dimensions.

Since $\bar{y}$ and $\bar{S}_m$ are independent from each other and $\bar{S}_m$ follows an i.i.d., the probability of receiving $\bar{y}$ given that $\bar{S}_m$ has been transmitted can be calculated as follows:

$$\Pr(\bar{y}|\bar{S}_m) = \prod_{i=1}^{N} \Pr(\bar{y}_i|\bar{S}_{m,i}) \tag{10.3}$$

$\Pr(\bar{y}_i|\bar{S}_{m,i})$ in Equation (10.3) is the conditional probability of receiving the ith time sample of $\bar{y}$, given ith time sample of tag response $\bar{S}_m$. $\Pr(\bar{y}_i|\bar{S}_{m,i})$ can be calculated using the well-known probability density function (pdf) of Gaussian distribution as shown in Equation (10.4).

$$\Pr(\bar{y}_i|\bar{S}_{m,i}) = \frac{1}{\sqrt{2\pi\sigma_\omega^2}}\ \exp\left(-\frac{1}{2}(y_i - hS_{m,i})\frac{1}{\sigma_\omega^2}\ (y_i - hS_{m,i})\right) \tag{10.4}$$

Using Equations (10.3) and (10.4), $\Pr(\bar{y}|\bar{S}_m)$ can be calculated as follows:

$$\Pr(\bar{y}|\bar{S}_m) = \frac{1}{\sqrt{2\pi\sigma_\omega^2}}\ \exp\left(\sum_{i=1}^{N}\left(-\frac{1}{2}(y_i - hS_{m,i})\frac{1}{\sigma_\omega^2}\ (y_i - hS_{m,i})\right)\right) \tag{10.5}$$

Equation (10.5) can be represented in vectorized form as below. $[.]^T$ is the Hermitian transpose.

$$\Pr(\bar{y}|\bar{S}_m) = \frac{1}{\sqrt{2\pi\sigma_\omega^2}}\ \exp\left(-\frac{1}{2}(\bar{y} - h\bar{S}_m)\frac{1}{\sigma_\omega^2}\ (\bar{y} - h\bar{S}_m)^T\right) \tag{10.6}$$

The received signal, $\bar{y}$ and all the tag combinations, $\bar{S}_m$ are used to evaluate probabilities given by Equation (10.6). The tag combination $\bar{S}_m$ producing highest probability is selected to be the detected tag, m. Therefore, $\Pr(\bar{y}|\bar{S}_m)$ is maximized over all possible $\bar{S}_m$ combinations for tag detection as follows:

$$\max_{\bar{S}_m}\ \Pr(\bar{y}|\bar{S}_m) \tag{10.7}$$

In Equation (10.6), only the $\exp$ (.) component is varying with $\bar{S}_m$. Hence, the detector proposed for this model simplifies to Equation (10.8).

$$\max_{\bar{S}_m} \Pr(\bar{y}|\bar{S}_m) = \min_{\bar{S}_m} \left((\bar{y} - h\bar{S}_m)(\bar{y} - h\bar{S}_m)^T \right) \tag{10.8}$$

The optimization given by Equation (10.8) performs the tag detection. Under these assumptions, the detector for the proposed signal model is the same as the minimum distance detector. Tag detector used in this section assumes the perfect channel knowledge and both the channel and received signals are considered to be real, even though in reality a typical RFID reader may perform I/Q demodulation hence dealing with complex signals. We release that assumption on next sections by allowing signals to be having both real and imaginary components.

The proposed signal models under different scenarios is listed in Figure 10.3. *Signal model II* still assumes perfect channel knowledge; however, it utilizes the information available in both the amplitude and phase for decision making. *Signal model III* needs to know only the statistical properties of the channel, while the actual channel realization is not required for decision making. In *signal model IV*, the channel is assumed to be unknown and a joint optimization on both the channel and tag type detection is performed. *Signal V* is derived for an existing chipless RFID system that utilizes only the power magnitudes of the backscattered tag response.

10.2.2   System Model II–Complex Signals

The signal model considered in this section is very similar to the *System Model I* discussed in the previous section. The same assumptions in *System Model I* apply; however, both the channels and all signals are treated as complex signals. Therefore, I/Q demodulation is implemented at the RFID reader, which is the case for some of the existing RFID readers. As a result, the readers no longer need very high sampling rates and only sample the baseband signals for I & Q. For such a reader, the received signal is called $\bar{y}$ and can be represented using Equation (10.9)

$$\bar{y} = h\bar{S}_m + \bar{\omega} \tag{10.9}$$

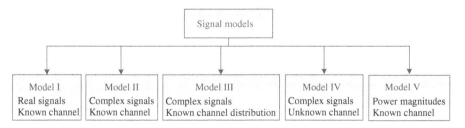

Figure 10.3   Proposed signal models.

The product of the forward and reverse channels h is assumed to be known and the signals considered in this model are all complex numbers. They can be represented using real and imaginary quantities for computation simplicity as follows:

$$h = h_r + jh_i$$
$$\bar{\omega} = \bar{\omega}_r + j\bar{\omega}_i$$

Therefore, the received signal can be written as

$$\bar{y} = \bar{y}_r + j\bar{y}_i$$
$$\bar{y}_r = h_r \bar{S}_{m,r} - h_i \bar{S}_{m,i} + \bar{\omega}_r \qquad (10.10)$$
$$\bar{y}_i = h_r \bar{S}_{m,i} + h_i \bar{S}_{m,r} + \bar{\omega}_i$$

The noise $\bar{\omega}$ added at the reader can be assumed to be independent of the filter response $\bar{S}_m$. In addition, real and imaginary components of individual time samples of $\bar{\omega}$ vector are assumed to follow an independent and identical Gaussian distribution (i.i.d.) with zero mean and a variance of σ_ω^2.

$$\bar{\omega}_r \sim N\,(\bar{0}, \sigma_\omega^2 I_N)$$
$$\bar{\omega}_i \sim N\,(\bar{0}, \sigma_\omega^2 I_N) \qquad (10.11)$$

A new vector, $\bar{y}_0$ having only real values is created using $\bar{y}_r$ and $\bar{y}_i$ as follows:

$$\bar{y}_0 = (\bar{y}_r\,, \bar{y}_i) \qquad (10.12)$$

Mean and covariance of $\bar{y}_0$ can be calculated as follows:

$$E[\bar{y}_0] = \bar{\mu} = [h_r \bar{S}_{m,r} - h_i \bar{S}_{m,i}\,, h_r \bar{S}_{m,i} + h_i \bar{S}_{m,r}]$$
$$\mathrm{Cov}[\bar{y}_0] = E[(\bar{y}_0 - \bar{\mu})^T (\bar{y}_0 - \bar{\mu})] \qquad (10.13)$$
$$= \sigma_\omega^2\, I_{2N}$$

I_{2N} in Equation (10.13) is the identity matrix with a dimension of $2N \times 2N$. Using the statistical properties calculated in Equation (10.13) the distribution of the real vector, $\bar{y}_0$ can be represented as follows:

$$\bar{y}_0 \sim N\,(\bar{\mu}, \sigma_\omega^2\, I_{2N})$$

Similar to the previous sections, the probability of receiving $\bar{y}_0$ given that $\bar{S}_m$ has been transmitted can be calculated as follows:

$$\Pr(\bar{y}_0 | \bar{S}_m) = \frac{1}{2\pi \sqrt{|\mathrm{Cov}[\bar{y}_0]|}} \quad \exp\left(-\frac{1}{2}(\bar{y}_0 - \bar{\mu})\,\mathrm{Cov}(\bar{y}_0)^{-1}\,(\bar{y}_0 - \bar{\mu})^T\right)$$
$$= \frac{1}{2\pi\sigma_\omega} \quad \exp\left(-\frac{1}{2\sigma_\omega^2}(\bar{y}_0 - \bar{\mu})\,(\bar{y}_0 - \bar{\mu})^T\right)$$

$$(10.14)$$

Similar to the previous detectors, Equation (10.14) is evaluated for all the possible tag combinations and the one with the highest probability is taken as the detector output. However, it can be seen that the detector can be further simplified to minimizing the exp (.) component. Therefore, the objective function of the detector can be represented as follows:

$$\max_{\bar{S}_m} \ \Pr(\bar{y}_0 | \bar{S}_m) = \min_{\bar{S}_m} \ \left((\bar{y}_0 - \bar{\mu}) \, (\bar{y}_0 - \bar{\mu})^T \right) \qquad (10.15)$$

Under the assumptions followed for the proposed signal model, the optimum detector is the same as the minimum distance detector. However, $\bar{y}_0$ and $\bar{\mu}$ can be calculated using Equations (10.12) and (10.13), respectively. Tag detector used in this section assumes the perfect channel knowledge and both the channel and received signals are considered to be complex, which means in reality the RFID reader performs I/Q modulation/demodulation. However, the expression in Equation (10.15) needs only real number calculations, hence lowers the computation complexity. In the following section, we assume perfect channel knowledge is no longer available. However, statistical properties of the channel are assumed to be available while I/Q modulation/demodulation is assumed to be performed at the RFID reader.

10.2.3   System Model III - Channel with a Known Distribution

The signal model discussed here assumed that the channel is no longer known. However, the statistical properties of the product of forward and reverse channels are approximated by a Gaussian distribution with a known mean and a variance. Similar to previous models, the channel is assumed to be constant throughout each tag reading. In addition, both the channel and the signals considered in this model are assumed to be complex, meaning I/Q modulation is performed at the RFID reader. Then the received signal can be modeled as

$$\bar{y} = h\bar{S}_m + \bar{\omega}$$

For computation simplicity, the complex channel (h) and the noise ($\bar{\omega}$) can be represented using two real components as follows:

$$h = h_r + jh_i$$
$$\bar{\omega} = \bar{\omega}_r + j\bar{\omega}_i$$

Similar to previous signal models, $\bar{S}_m$ is the signal transmitted by the tag and $\bar{\omega}$ is the noise added at the receiver that is independent of $\bar{S}_m$ and each noise sample follows an independent and identical Gaussian distribution. The statistical properties of the real and imaginary components of the noise ($\bar{\omega}_r$ and $\bar{\omega}_i$) is given by Equation (10.16).

$$\bar{\omega}_r \sim N\left(\bar{0}, \sigma_\omega^2 I_N\right)$$
$$\bar{\omega}_i \sim N\left(\bar{0}, \sigma_\omega^2 I_N\right) \qquad (10.16)$$

It is assumed that both the real and imaginary components of noise are having the same statistical properties. Next, the product of forward and reverse channels h is assumed to have Gaussian distributions for the real and imaginary components as shown in Equation (10.17).

$$h_r \sim N\ (\mu_r, \sigma_r^2)$$
$$h_i \sim N\ (\mu_i, \sigma_i^2)$$

(10.17)

Then the real and imaginary components of the received signal $\bar{y}$ can be represented using the following relationship:

$$\bar{y} = \bar{y}_r + j\bar{y}_i$$
$$\bar{y}_r = h_r \bar{S}_{m,r} - h_i \bar{S}_{m,i} + \bar{\omega}_r$$
$$\bar{y}_i = h_r \bar{S}_{m,i} + h_i \bar{S}_{m,r} + \bar{\omega}_i$$

(10.18)

A new real vector, $\bar{y}_0$ is created using $\bar{y}_r$ and $\bar{y}_i$ as follows:

$$\bar{y}_0 = [\bar{y}_r, \bar{y}_i]$$

(10.19)

The statistical properties of $\bar{y}_r$ and $\bar{y}_i$ are examined next. It can easily be seen that they too follow a Gaussian distribution. The mean of $\bar{y}_r$ and $\bar{y}_i$ is given by Equation (10.20).

$$E[\bar{y}_r] = \mu_r \bar{S}_{m,r} - \mu_i \bar{S}_{m,i}$$
$$E[\bar{y}_i] = \mu_r \bar{S}_{m,i} + \mu_i \bar{S}_{m,r}$$

(10.20)

Covariances of $\bar{y}_r$ and $\bar{y}_i$ can be calculated using the following formula:

$$\mathrm{Cov}[\bar{X}] = E[(\bar{X} - E[\bar{X}])^T\ (\bar{X} - E[\bar{X}])]$$

After some calculations, it can be shown that the covariances of $\bar{y}_r$ and $\bar{y}_i$ are as follows:

$$\mathrm{Cov}(\bar{y}_r) = \sigma_r^2 \bar{S}_{m,r}^T \bar{S}_{m,r} + \sigma_i^2 \bar{S}_{m,i}^T \bar{S}_{m,i} + \sigma_\omega^2 I_N$$
$$\mathrm{Cov}(\bar{y}_i) = \sigma_r^2 \bar{S}_{m,i}^T \bar{S}_{m,i} + \sigma_i^2 \bar{S}_{m,r}^T \bar{S}_{m,r} + \sigma_\omega^2 I_N$$

(10.21)

Therefore, the distribution of $\bar{y}_r$ and $\bar{y}_i$ can be listed as follows:

$$\bar{y}_r \sim N\ (\mu_r \bar{S}_{m,r} - \mu_i \bar{S}_{m,i}\ ,\ \sigma_r^2 \bar{S}_{m,r}^T \bar{S}_{m,r} + \sigma_i^2 \bar{S}_{m,i}^T \bar{S}_{m,i} + \sigma_\omega^2 I_N$$
$$\bar{y}_i \sim N\ (\mu_r \bar{S}_{m,i} + \mu_i \bar{S}_{m,r}\ ,\ \sigma_r^2 \bar{S}_{m,i}^T \bar{S}_{m,i} + \sigma_i^2 \bar{S}_{m,r}^T \bar{S}_{m,r} + \sigma_\omega^2 I_N$$

(10.22)

Using Equations (10.19) and (10.22), it can be concluded that $\bar{y}_0$ has a multivariate Gaussian distribution with a dimension of 2. The mean of $\bar{y}_0$ can be written as

$$E[\bar{y}_0] = [\mu_r \bar{S}_{m,r} - \mu_i \bar{S}_{m,i}\ ,\ \mu_r \bar{S}_{m,i} + \mu_i \bar{S}_{m,r}]$$

(10.23)

The covariance of $\bar{y}_0$ can be calculated as follows:

$$\text{Cov}[\bar{y}_0] = E[(\bar{y}_0 - E[\bar{y}_0])^T \ (\bar{y}_0 - E[\bar{y}_0])]$$

After some calculations, it can be seen that $\text{Cov}[\bar{y}_0]$ simplifies to

$$\text{Cov}[\bar{y}_0] = \begin{pmatrix} \sigma_r^2 \bar{S}_{m,r}^T \bar{S}_{m,r} + \sigma_i^2 \bar{S}_{m,i}^T \bar{S}_{m,i} & \sigma_r^2 \bar{S}_{m,r}^T \bar{S}_{m,i} - \sigma_i^2 \bar{S}_{m,i}^T \bar{S}_{m,r} \\ \sigma_r^2 \bar{S}_{m,i}^T \bar{S}_{m,r} - \sigma_i^2 \bar{S}_{m,r}^T \bar{S}_{m,i} & \sigma_r^2 \bar{S}_{m,i}^T \bar{S}_{m,i} + \sigma_i^2 \bar{S}_{m,r}^T \bar{S}_{m,r} \end{pmatrix}$$

$$+ \sigma_\omega^2 \ I_{2N} \tag{10.24}$$

Then the conditional probability on receiving $\bar{y}_0$ given that $\bar{S}_m$ has been transmitted is given by Equation (10.25):

$$\text{Pr}(\bar{y}_0|\bar{S}_m) = \frac{1}{2\pi \sqrt{|\text{Cov}[\bar{y}_0]|}} \exp\left(-\frac{1}{2}(\bar{y}_0 - E[\bar{y}_0]) \ \text{Cov}(\bar{y}_0)^{-1} \ (\bar{y}_0 - E[\bar{y}_0])^T\right) \tag{10.25}$$

Similar to previous models, now the probability calculated in Equation (10.25) is maximized over all possible tag combinations, $\bar{S}_m$:

$$\max_{\bar{S}_m} \ \text{Pr}(\bar{y}_0|\bar{S}_m) \tag{10.26}$$

In this model, the product of the forward and reverse channels is modeled using a Gaussian distribution. In the next model, we assume no channel information is available. As a result, it is a joint optimization problem of deciding both the channel and the tag combination.

10.2.4 System Model IV–Unknown Channel

In this model, we assume no channel knowledge is available to the RFID reader. In addition, signals considered here are complex, meaning I/Q modulation/ demodulation is utilized at the reader. It is assumed that the product of the forward and reverse channels, h, is an unknown complex number and is a constant during the interrogation time. Then the received signal $\bar{y}$ can be written as

$$\bar{y} = h\bar{S}_m + \bar{\omega} \tag{10.27}$$

For computation simplicity, the complex channel and the noise can be represented using two real components as follows:

$$h = h_r + jh_i$$
$$\bar{\omega} = \bar{\omega}_r + j\bar{\omega}_i$$

Then the received signal $\bar{y}$ can be represented using the real and imaginary components similar to previous models.

$$\bar{y} = \bar{y}_r + j\bar{y}_i$$
$$\bar{y}_r = h_r \bar{S}_{m,r} - h_i \bar{S}_{m,i} + \bar{\omega}_r \tag{10.28}$$
$$\bar{y}_i = h_r \bar{S}_{m,i} + h_i \bar{S}_{m,r} + \bar{\omega}_i$$

From Equation (10.28), it can be clearly seen that conditional probability of receiving $\bar{y}_r$ and $\bar{y}_i$ given $\bar{S}_m$ and h has a Gaussian distribution. Their statistical properties can be calculated as follows:

$$E[\bar{y}_r | h, \bar{S}_m] = h_r \bar{S}_{m,r} - h_i \bar{S}_{m,i}$$
$$E[\bar{y}_i | h, \bar{S}_m] = h_r \bar{S}_{m,i} + h_i \bar{S}_{m,r}$$
$$\text{Cov}[\bar{y}_r | h, \bar{S}_m] = \sigma_\omega^2 I_N \tag{10.29}$$
$$\text{Cov}[\bar{y}_i | h, \bar{S}_m] = \sigma_\omega^2 I_N$$

A vector ($\bar{y}_0$) containing real values is created by stacking $\bar{y}_r$ and $\bar{y}_i$ on a row vector as follows:

$$\bar{y}_0 = [\bar{y}_r , \bar{y}_i] \tag{10.30}$$

Similar to $\bar{y}_r$ and $\bar{y}_i$, when both the channel h and the tag combination $\bar{S}_m$ are given, the conditional probability of receiving $\bar{y}_0$ is having a Gaussian distribution. The mean is given by

$$E[\bar{y}_r | h, \bar{S}_m] = \bar{\mu} = [h_r \bar{S}_{m,r} - h_i \bar{S}_{m,i} , h_r \bar{S}_{m,i} + h_i \bar{S}_{m,r}] \tag{10.31}$$

In order to derive an expression for the conditional probability, covariance has to be calculated. The following formulas show how to calculate the covariance:

$$\text{Cov}[\bar{y}_0 | h, \bar{S}_m] = E[[\bar{y}_0 - \bar{\mu}]^T [\bar{y}_0 - \bar{\mu}]]$$
$$= E[[\bar{\omega}_r , \bar{\omega}_i]^T [\bar{\omega}_r , \bar{\omega}_i]]$$
$$= \sigma_\omega^2 I_{2N}$$

Then the conditional probability of receiving $\bar{y}_0$ given h and $\bar{S}_m$ can be calculated as in Equation (10.32):

$$\bar{y}_0 | h, \bar{S}_m \sim N(\bar{\mu} , \sigma_\omega^2 I_{2N})$$
$$\Pr(\bar{y}_0 | h, \bar{S}_m) = \frac{1}{2\pi\sigma_\omega} \exp\left(-\frac{1}{2\sigma_\omega^2}(\bar{y}_0 - \bar{\mu})^T (\bar{y}_0 - \bar{\mu})\right) \tag{10.32}$$

The probability given in Equation (10.32) is maximized over all possible combinations of h and $\bar{S}_m$. Therefore, this is a joint optimization problem.

$$\max_{h_r,h_i,\bar{S}_m} \Pr(\bar{y}_0|h_r,h_i,\bar{S}_m) = \min_{h_r,h_i,\bar{S}_m} ((\bar{y}_0 - \bar{\mu})\,(\bar{y}_0 - \bar{\mu})^T)$$

$$= \min_{h_r,h_i,\bar{S}_m} (\bar{y}_0\bar{y}_0^T - 2\bar{y}_0\bar{\mu}^T + \bar{\mu}\bar{\mu}^T) \qquad (10.33)$$

10.2.5   Joint Optimization of h and Tag Type

$\bar{\mu}$ is calculated using Equation (10.31). However, there are infinitely large number of combinations for h_r and h_i; hence, it is not computationally feasible. A feasible solution would be to first find the optimum channel for a given tag combination. Then the given tag combination response and the optimum channel are used for calculating the conditional probability given in Equation (10.32). Then the same process is repeated for all possible tag combinations, similar to previous detectors, to calculate the highest probability. Next, calculating the optimum channel for a given tag combination is discussed. $L(h_r, h_i)$ is defined as follows:

$$L(h_r, h_i) = \bar{y}_0\bar{y}_0^T - 2\bar{y}_0\bar{\mu}^T + \bar{\mu}\bar{\mu}^T \qquad (10.34)$$

For optimum h_r and h_i, following conditions have to be satisfied:

$$\frac{\partial}{\partial h_r}L(h_r, h_i) = 0$$
$$\frac{\partial}{\partial h_i}L(h_r, h_i) = 0 \qquad (10.35)$$

Using Equation (10.31), it can be shown that

$$\frac{\partial\bar{\mu}}{\partial h_r} = [\bar{S}_{m,r}\,,\bar{S}_{m,i}]$$
$$\frac{\partial\bar{\mu}}{\partial h_i} = [-\bar{S}_{m,i}\,,\bar{S}_{m,r}] \qquad (10.36)$$

Using the relationships in Equations (10.34)–(10.36), the optimum channel $\hbar_r$ and $\hbar_i$ for a given $\bar{S}_m$ can be derived as follows:

$$\hbar_r = \frac{\bar{y}_0[\bar{S}_{m,r}\,,\bar{S}_{m,i}]^T}{[\bar{S}_{m,r}\bar{S}_{m,r}^T + \bar{S}_{m,i}\bar{S}_{m,i}^T]}$$
$$\hbar_i = \frac{\bar{y}_0[-\bar{S}_{m,i}\,,\bar{S}_{m,r}]^T}{[\bar{S}_{m,r}\bar{S}_{m,r}^T + \bar{S}_{m,i}\bar{S}_{m,i}^T]} \qquad (10.37)$$

The optimum channel estimates obtained from Equation (10.37) are used to calculate the optimum $\bar{\mu}$ ($\bar{\mu}_0$). Equations (10.31) and (10.37) yield

$$\bar{\mu}_0 = [\hbar_r \bar{S}_{m,r} - \hbar_i \bar{S}_{m,i} \ , \ \hbar_r \bar{S}_{m,i} + \hbar_i \bar{S}_{m,r}] \tag{10.38}$$

Then, the new optimization problem reduces to

$$\max_{\hbar_r, \hbar_i, \bar{S}_m} \Pr(\bar{y}_0 | \hbar_r, \hbar_i, \bar{S}_m) = \min_{\hbar_r, \hbar_i, \bar{S}_m} ((\bar{y}_0 - \bar{\mu}_0)(\bar{y}_0 - \bar{\mu}_0)^T) \tag{10.39}$$

In this model, no channel information is available to the reader, and the only assumption is that the channel is static during the short interrogation time period. The tag detector derived in the model uses the received signal at the RFID reader and all the possible tag responses to determine both the channel and the tag combination that provides the highest probability.

All the four signal models discussed in this section are based on time-domain signal samples [1]. Some of the existing RFID readers work based on frequency-domain samples. The following section discusses the tag detectors that can work based on frequency-domain signal samples.

10.3 SYSTEM MODELS–FREQUENCY DOMAIN

Chipless RFID readers are based on either time-domain tags [2–4] or frequency-domain tags [5–11]. Time-domain tags encode the information in time samples of the signal leaving a unique time signature, whereas the frequency-domain tags encode information in the frequency samples of the signal leaving a unique frequency-domain signature. Therefore, it is important to examine the tag detection techniques for the frequency-domain-based chipless RFID tags. There is another very important benefit of using frequency-domain tags. Tag detectors derived for time-domain-based tags have a high computational complexity. However, frequency-domain-based tag detection can be achieved with relatively a lower computational complexity as explained in *Chapter 11* in detail. The rest of this section describes five tag detection techniques for frequency-based chipless RFID tags.

10.3.1 System Models I–IV

The tag detectors derived for time-domain chipless RFID tags are first revisited briefly. The signal model used in all the four detectors is as follows:

$$\bar{y}(t) = h\bar{S}_m(t) + \bar{\omega}(t) \tag{10.40}$$

10.3.1.1   *Model I*   In model I, channel h is a known real constant. Noise is having a zero mean normal distribution with a covariance $\sigma_{\bar{w}}^2 I_N$. Then the probability distribution function of receiving $\bar{y}$ given that $\bar{S}_m$ has been transmitted is given by Equation (10.6). If Fourier transformation is performed on Equation (10.40), the result is shown as follows:

$$\bar{Y}(f) = h\bar{S}_m(f) + \bar{w}(f) \tag{10.41}$$

$\bar{Y}(f)$ is the Fourier transform of the received signal $\bar{y}$. $\bar{S}_m(f)$ and $\bar{w}(f)$ are the Fourier transforms of the mth tag response $(\bar{S}_m)$ and the noise $\bar{w}$, respectively. Fourier transform is a unitary transformation. Therefore, the statistical properties of the signals should remain the same. As a result, the probability distribution function of receiving $\bar{y}(f)$ given that $\bar{S}_m(f)$ has been transmitted is the same as Equation (10.6). Therefore, the frequency-domain-based chipless RFID tag detector for system model I can be derived using the frequency samples of the signal as follows:

$$\max_{\bar{S}_m(f)} \Pr(\bar{Y}(f)|\bar{S}_m(f)) = \min_{\bar{S}_m(f)} \left((\bar{Y}(f) - h\bar{S}_m(f) \, (\bar{Y}(f) - h\bar{S}_m(f))^T \right) \tag{10.42}$$

10.3.1.2   *Model II*   Similarly, frequency-domain chipless RFID tag detector for system model II can be derived as follows:

$$\max_{\bar{S}_m(f)} \Pr(\bar{Y}_0(f)|\bar{S}_m(f) = \min_{\bar{S}_m(f)} \left((\bar{Y}_0(f) - \bar{\mu}(f)) \, (\bar{Y}_0(f) - \bar{\mu}(f)^T \right) \tag{10.43}$$

Similar to the previous model, $\bar{S}_m(f)$ is the Fourier transformation of $\bar{S}_m$. $\bar{Y}_0(f)$ and $\bar{\mu}(f)$ are the Fourier transformations of $\bar{y}_0$ in Equation (10.12) and $\bar{\mu}$ in Equation (10.13), respectively.

10.3.1.3   *Model III*   The statistical properties of the signal model III too remain the same, hence the tag detector in frequency domain is given by

$$\max_{\bar{S}_m(f)} \Pr(\bar{Y}_0(f)|\bar{S}_m(f)) = \frac{1}{2\pi \sqrt{|\mathrm{Cov}[\bar{Y}_0(f)]|}}$$

$$\exp\left(-\frac{1}{2}(\bar{Y}_0(f) - E[\bar{Y}_0(f)]) \, \mathrm{Cov}(\bar{Y}_0(f))^{-1} \right.$$

$$\left. (\bar{Y}_0(f) - E[\bar{Y}_0(f)])^T \right) \tag{10.44}$$

$E[\bar{Y}_0(f)]$ and $\mathrm{Cov}[\bar{Y}_0(f)]$ are the Fourier transformations of $E[\bar{y}_0]$ in Equation (10.23) and $\mathrm{Cov}[\bar{y}_0]$ in Equation (10.24), respectively.

10.3.1.4   Model IV   Similar to previous three models, statistical properties of the detector in signal model do not change with the Fourier transform. Therefore, the tag detector for signal model IV can be written as

$$
\max_{\hbar_r, \hbar_i, \bar{S}_m(f)} P(\bar{Y}_0(f)|\hbar_r, \hbar_i, \bar{S}_m(f))
$$

$$
= \min_{\hbar_r, \hbar_i, \bar{S}_m(f)} ((\bar{Y}_0(f) - \bar{\mu}_0(f)) \quad (\bar{Y}_0(f) - \bar{\mu}_0(f))^T) \tag{10.45}
$$

$\bar{Y}_0(f)$ and $\bar{\mu}_0(f)$ are the Fourier transformations of $\bar{y}_0(t)$ in Equation (10.30) and $\bar{\mu}_0(t)$ in Equation (10.38), respectively.

It is clear that the frequency samples of the signal can be used to detect tags using the same detectors derived for the time-based models. It is true that the frequency signature of the frequency-domain tags can be seen in Fourier transformation of the time-domain signal. However, there are chipless RFID readers that work on the power spectral density of the signal, rather than Fourier transformation. In the following section, a tag detector is derived for a power-based chipless RFID reading method.

10.3.2   System Model V – Power Magnitudes

There are chipless RFID readers that operate based on the power measurements rather than on voltage samples. In these readers, a narrow-banded sinusoidal waveform is transmitted as the interrogating signal and the magnitude of the tag response is compared with the transmitted signal using a gain detector. Then the frequency of the narrow-banded interrogating signal is swept across the frequency of interest and the frequency signature of the tag is obtained.

The frequency signature obtained in this method requires a lesser sampling rate at the reader compared to the frequency signature obtained using the Fourier transformation performed on time-domain-based measurements. In addition, the narrow-banded signals are subjected to lesser noise, which could lead to better tag reading reliability.

System Model V describes an existing chipless RFID reader developed at Monash Microwave, Antenna, RFID and Sensor Laboratory (MMARS) under Australian Research Council's Linkage Project Grant: *LP0991435: Backscatter-based RFID system capable of reading multiple chipless tags for regional and suburban libraries.* The reader works on power magnitude of the received tag response. The application assumes a fixed distance between the reader and the tags, and at short distances such as 15 cm, the line-of-sight component dominates over any multipaths. Therefore, the channel undergoes a very slow variation with respect to time. During the calibration phase, all possible tag responses are measured and recorded for the given distance between the reader and the tags. The recorded tag responses are used with the likelihood-based detector derived as follows.

If the Fourier transformation of the tag response is $\bar{S}_m(f)$ and the Fourier transformation of the noise added at the reader is $\bar{\omega}(f)$, then the Fourier transformation of the received signal at the reader is given by $\bar{Y}(f)$. Then $\bar{Y}(f)$ is separated into real and imaginary components as shown in Equation (10.46):

$$\bar{Y}(f) = \bar{S}_m(f) + \bar{w}(f)$$
$$\bar{Y}(f) = (\bar{S}_{m,r}(f) + \bar{w}_r(f)) + j(\bar{S}_{m,i}(f) + \bar{w}_i(f)) \tag{10.46}$$

As shown in Section 10.3.1, the statistical properties of $\bar{w}(f)$ is the same as its time-domain samples. Then the power magnitude Z of the frequency-domain samples are given by

$$Z = |\bar{Y}(f)|^2 = (\bar{S}_{m,r}(f) + \bar{w}_r(f))^2 + (\bar{S}_{m,i}(f) + \bar{w}_i(f))^2$$

Statistical properties of $\bar{w}_r(f)$ and $\bar{w}_i(f)$ can be written as follows:

$$\bar{w}_r(f) \sim N \ (\bar{0}, \sigma_r^2 I_N)$$
$$\bar{w}_i(f) \sim N \ (\bar{0}, \sigma_i^2 I_N)$$

Then the real (R) and imaginary (I) components of $\bar{Y}(f)$ are defined as follows:

$$R = \bar{S}_{m,r}(f) + \bar{w}_r(f)$$
$$I = \bar{S}_{m,i}(f) + \bar{w}_i(f)$$

The statistical properties of R and I can be derived as shown in Equation (10.47).

$$R \sim N \ (\bar{S}_{m,r}(f), \sigma_r^2 I_N)$$
$$I \sim N \ (\bar{S}_{m,i}(f), \sigma_i^2 I_N) \tag{10.47}$$

Assuming independence between individual samples of each R and I, the probability of receiving Z given that $\bar{S}_{m,r}(f)$ and $\bar{S}_{m,i}(f)$ had been transmitted is given by Equation (10.48). $(.)^{(k)}$ is the kth sample of the corresponding vector.

$$\Pr(Z|\bar{S}_{m,r}(f), \bar{S}_{m,i}(f)) = \prod_{k=1}^{N} \Pr(Z^{(k)}|S_{m,r}^{(k)}(f), S_{m,r}^{(k)}(f)) \tag{10.48}$$

It is clear that R and I are independent. As a result, Z has a noncentral chi-square distribution with λ being the noncentrality parameter and df being the degree of freedom, which is 2. λ for kth sample can be calculated as follows:

$$\lambda^{(k)} = \left[\frac{S_{m,r}^{(k)}(f)}{\sigma_r}\right]^2 + \left[\frac{S_{m,i}^{(k)}(f)}{\sigma_i}\right]^2 \tag{10.49}$$

Then $\Pr(Z^{(k)}|S^{(k)}_{m,r}(f), S^{(k)}_{m,r}(f))$ can be expressed as follows:

$$\Pr(Z^{(k)}|S^{(k)}_{m,r}(f), S^{(k)}_{m,r}(f)) = \frac{1}{2}\exp\left(-\frac{Z^{(k)} + \lambda^{(k)}}{2}\right)\left(\frac{Z^{(k)}}{\lambda^{(k)}}\right)^{\frac{df}{4}-\frac{1}{2}}$$
$$I_{df/2-1}(\sqrt{\lambda^{(k)} \cdot Z^{(k)}}) \tag{10.50}$$

$I_v(.)$ is the modified Bessel function of the first kind. The above expression can be simplified using $df = 2$ and assuming the variance for both real and imaginary noise components is the same ($\sigma_r = \sigma_i = \sigma_w$). Then,

$$\Pr(Z^{(k)}|S^{(k)}_{m,r}(f), S^{(k)}_{m,r}(f)) = \frac{1}{2}\exp\left(-\frac{Z^{(k)} + \lambda^{(k)}}{2}\right)I_0(\sqrt{\lambda^{(k)} \cdot Z^{(k)}}) \tag{10.51}$$

Using Equations (10.48) and (10.51), $\Pr(Z|\bar{S}_{m,r}(f), \bar{S}_{m,i}(f))$ can be calculated as follows:

$$\Pr(Z|\bar{S}_{m,r}(f), \bar{S}_{m,i}(f)) = \frac{1}{2^N}\exp\left(-\frac{1}{2}\left(N \quad E[Z] + E_m/\sigma_w^2\right)\right)$$
$$\prod_{k=1}^{N} I_0(\sqrt{\lambda^{(k)} Z^{(k)}}) \tag{10.52}$$

E_m in Equation (10.52) is the energy of the selected tag response $\bar{S}_m(f)$ defined as

$$E_m = \sum_{k=1}^{N}([S^{(k)}_{m,r}(f)]^2 + [S^{(k)}_{m,i}(f)]^2) \tag{10.53}$$

Then the probability calculated in Equation (10.52) is maximized over $\bar{S}_{m,r}(f)$ and $\bar{S}_{m,i}(f)$ to detect the most likelihood tag combination.

$$\max_{\bar{S}_{m,r}(f), \bar{S}_{m,i}(f)} \Pr(Z|\bar{S}_{m,r}(f), \bar{S}_{m,i}(f)) \tag{10.54}$$

The tag detection technique developed in this section is used with an existing chipless RFID reader. During the calibration phase, the tag responses for all possible combinations are recorded. Then, the above detection technique is used to detect a random tag.

So far in the chapter, several tag detection techniques have been derived based on both time-domain samples and frequency-domain samples. In order to test these detection techniques comprehensively, a MATLAB simulation is performed, which is explained in the following section.

10.4   SIMULATIONS

The validity of the above tag detection techniques is verified using MATLAB and Computer Simulation Technology (CST) simulations. The steps carried out in the simulation are given in Figure 10.4. First, an interrogating signal was generated to provide a flat frequency response in the 2.2–2.6 GHz frequency range. Four resonators were designed using CST with resonating frequencies as shown in Table 10.1. The resonance frequencies were selected as 100 MHz apart following the specifications provided in Ref. [5]. Then the combinations of resonators were placed beside a microstrip line to cover all possible tag IDs. One end of the microstrip line containing bandstop filters is fed with the interrogating signal and the tag responses are collected at the other end. These collected tag responses were saved in a lookup table for the algorithms to be used later. More details about the tag design are available in *Chapter 9*.

Then the tag responses were fed through a channel that is given by the product of the forward and reverse channels of the RFID system. Depending on the detection technique, the channel values were selected to be a known constant or a variable with a known or unknown statistical distribution. Finally, noise was added to the resultant signal according to the specified signal-to-noise ratio (SNR).

SNR was calculated compared to the average power of all the tag combinations. It can be summarized as follows:

$I(t)$ – interrogating signal

h_f – forward channel

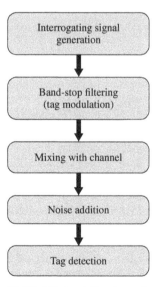

Figure 10.4   Flowchart of the MATLAB simulation in conjunction with CST full-wave EM solver simulation.

TABLE 10.1　Simulation Parameters

Parameter	Value
Center frequency	2.4 GHz
Total bits encoded in a tag	4 bits
Flat frequency response	400 MHz
Bandstop filter attenuation	10 dB
Bandstop filter 3 dB bandwidth	40 MHz
Guard-band	60 MHz
Resonance frequency set 1 (MSB to LSB)	[2.2, 2.3, 2.4, 2.5] GHz
Resonance frequency set 2 (MSB to LSB)	[2.34, 2.38, 2.42, 2.46] GHz
Resonance frequency set 3 (MSB to LSB)	[22.5, 23.5, 24.5, 25.5] GHz
Channel mean	0.4
Channel standard deviation	0.1
Number of iterations	Up to 10,000,000

h_r – reverse channel
h – product of forward and reverse channels
$F_m(t)$ – impulse response of the mth filter
$S_m(t)$ – mth tag response
$y(t)$ – received signal
$\omega(t)$ – noise added at the reader.

Then the received signal can be represented using Equation (10.55).

$$
\begin{aligned}
y(t) &= [[h_f I(t)] * F_m(t)]h_r + \omega(t) \\
&= h_f h_r[I(t) * F_m(t)] + \omega(t) \\
&= hS_m(t) + w(t)
\end{aligned}
\tag{10.55}
$$

The power of each tag response was calculated and averaged to obtain the average power of a given tag response. For example, 2-bit tags have four different tag responses $(S_m(t))$ and power of each tag response is calculated and averaged to obtain the average power of 2-bit tag responses. Then the average tag response power was multiplied using the channel to calculate the average signal power available at the reader. For a given SNR, noise power was calculated using this available signal power.

MATLAB simulation parameters are outlined in 10.1. Four bandstop filters are used, and most significant bit (MSB) corresponds to the lowest resonance frequency and least significant bit (LSB) to the highest.

I/Q demodulation was performed with the received signal at the RFID reader. Then, the two output time-domain signal vectors were used to evaluate likelihood expressions for each detection technique. In order to verify their frequency-domain performances the time-domain vectors were converted using fast Fourier transform (FFT). Then, these frequency-domain samples were used for tag detection.

Finally, the DER is calculated for each tag detection technique at different SNR levels. DER is defined as the probability of having at least one erroneous bit out of all data bits. It can be represented using the throughput as shown in Equation (10.56). Throughput is defined as the ratio between the number of successful tag readings (N_S) and the total number of tag readings performed (N_T) as given in Equation (10.57).

$$DER = 1 - throughput \qquad (10.56)$$

$$Throughput = \frac{N_S}{N_T} \qquad (10.57)$$

Existing chipless RFID systems use a threshold-based detection technique based on frequency-domain-based samples. The presence and absence of each resonator in this method are detected based on a magnitude threshold at corresponding resonator frequency bands. This threshold-based detection technique is implemented and the DER is calculated to compare the performances of the proposed detection techniques.

10.5 EXPERIMENTAL SETUP

The likelihood detector derived under this *System model V* uses only the magnitude of the tag responses in the frequency domain. This tag detection technique is derived for an existing chipless RFID system that operates between 21 and 27 GHz. A circular resonator-based chipless RFID tags are designed and fabricated as described in *Chapter 9*. A printed tag encoded with bits [1111] is shown in Figure 10.5.

Two tag prototypes were experimented to validate the theory. Tag type I is a retransmission tag operating over the frequency band of 2.2–2.6 GHz. The tag

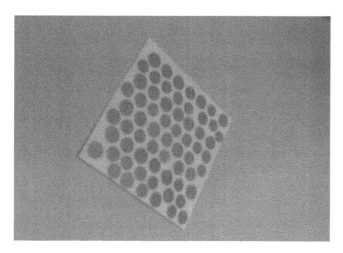

Figure 10.5 A chipless tag coded with bits [1111].

Tag   Reader RF electronics   Digital control and
                                                                power supply unit

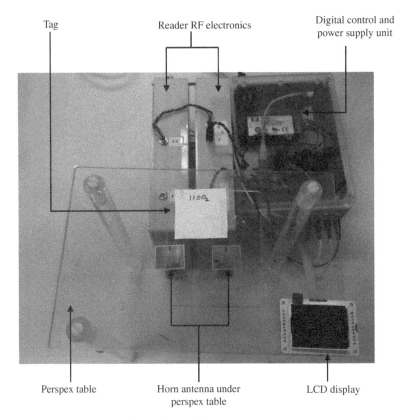

Perspex table   Horn antenna under   LCD display
                            perspex table

Figure 10.6   Experimental setup.

type II is a backscattering tag operating at 21–27 GHz. The experimental setup
for validating tag type II is shown in Figure 10.6. In both experiments, the reader
transmits a narrow-banded sinusoid signal using one horn antenna and the reader
receives the tag response using a second antenna. The magnitude of the received
signal is recorded along with the frequency of the signal sent. The frequency is
swept across the 21–27 GHz band and the complete band response is recorded.

This process is repeated with all the possible 16 combinations for a 4-bit tag and the
recorded tag responses are used with the tag detection technique derived for *System
model V* to detect the encoded tag data bits.

In the following section, both simulation and experimental results are analyzed for
the five tag detection techniques presented.

10.6   RESULTS

An interrogating signal is designed to provide a flat frequency response for at
least 400 MHz around 2.4 GHz and it is used as the port excitation signal in CST

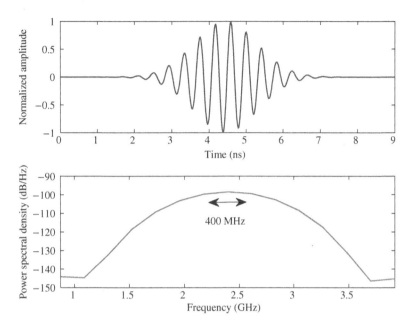

Figure 10.7 Interrogating signal in time and frequency domain.

simulations. Figure 10.7 shows the time- and frequency-domain interrogating signals. It can clearly be seen that the above requirement is achieved quite easily.

Then the multiresonator tags are designed in CST according to the specifications given in Table 10.1. Initially, resonators are designed according to the frequency set 1 given in Table 10.1 that includes a guard-band between resonance frequencies. Then the resonators are redesigned in CST using the frequency set 2 without any guard-band. Then both time- and frequency-domain tag responses are obtained from CST simulations and stored in a lookup table for MATLAB simulations.

Figure 10.8 shows the frequency-domain response of a tag encoded with bits [1111]. It can clearly be observed that the resonances occur at the designed frequencies given by set 1 in Table 10.1. Figure 10.8 also shows the corresponding time-domain response of a tag encoded with bits [1111].

Then a new set of tag resonators are designed according to the frequency set 2 given in Table 10.1 without any guard-band. The frequency- and time-domain tag responses are illustrated in Figure 10.9. Similar to the previous simulation results, it can be concluded that the resonators are performing as expected by observing the resonance frequencies.

Then these four resonators are arranged to implement the 16 tag combinations for a 4-bit tag. Simulations are repeated for 16 times and both time- and frequency-domain tag responses are obtained and recorded. Next the system models discussed in the previous section are implemented in MATLAB and the performances are analyzed. DERs under different SNR levels are calculated for each system model and the compared with a threshold-based tag detection system used

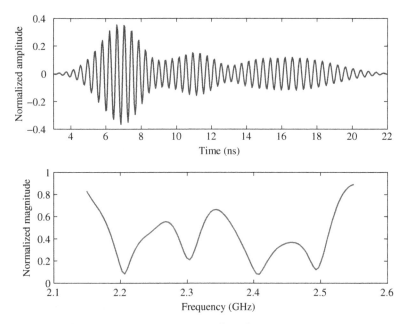

Figure 10.8   Tag responses for [1111] with a guard-band.

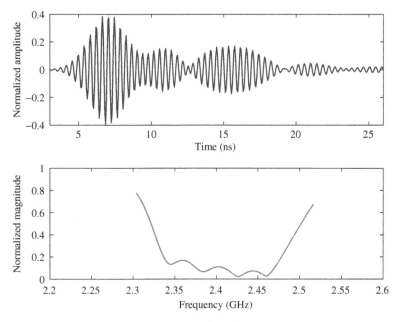

Figure 10.9   Tag responses for [1111] without a guard-band.

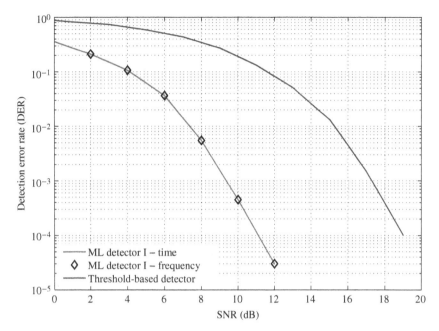

Figure 10.10 DER versus SNR for 4-bit tag with 60 MHz guard-band.

in existing RFID readers. Simulation results for each system model are presented in the following sections.

10.6.1 System Model I

First, the DER is calculated at different noise power levels, effectively changing SIR and the result for *System Model I* is shown in Figure 10.10.

It can be seen that the DER is the same for both time- and frequency-domain-based samples. Therefore, it verifies the argument that the time-domain-based detection expression remains valid for the frequency-domain-based samples. When the frequency-domain-based samples are used, tag detection can be performed using the presence and absence of the power dips at resonating frequencies. This is achieved in existing chipless RFID readers using a threshold-based detection method that detects the power dips. This threshold-based detection method is used as the baseline comparison for the proposed tag detection methods.

As can be seen in 10.10, the proposed detection methods provide a significant improvement on DER over the threshold-based detection method. It can be seen that in order to achieve 99.99% reading accuracy, threshold-based detector needs an SNR of 19 dB. However, the proposed detector requires an SNR of only 11 dB, which is an SNR gain of 8 dB over the threshold-based detector. SNR can be related to the reading distance and SNR gain results in an increment in the tag reading range. The signal travels twice the distance between the reader and the tag. As a result, if the indoor propagation constant is assumed to be 2 [12], this SNR improvement can be related to improve the reader distance by a factor of 2.5. Therefore, the tag reading range can be improved by 2.5 times with the proposed tag detection technique while achieving a target reading accuracy of 99.99%.

On the other hand, this improved performance can be viewed as an increment in the tag reading accuracy at a given SNR. For example, the DER of the threshold-based detector at SNR = 10 dB is about 90%. With the proposed detector, the accuracy can be improved up to 99.95%. This avoids the requirement to perform multiple tag readings to detect one tag, especially under low-SNR scenarios. Therefore, depending on the application, improvement can be viewed on either the tag reading accuracy or the tag reading range.

Then the *System Model I* is used to detect the tag responses obtained using frequency set 2 given in Table 10.1. The removal of the guard-band causes the number of resonators allowed per unit bandwidth to be more than doubled in this example, which in turn double the tag bit capacity. As can be seen in Figure 10.11, traditional threshold-based decoder produces very poor performances when the guard-band is absent. For example, at an SNR of 10 dB the DER of the threshold-based method is about 60%. However, the proposed *Signal Model I* detector still achieves a high accuracy of 99% at the same SNR.

Similar to the increment in the tag reading accuracy, the SNR gain of the proposed method compared to the baseline is also significant. For example, in order to achieve an accuracy of 90%, threshold-based detector requires at least an SNR of 19 dB while

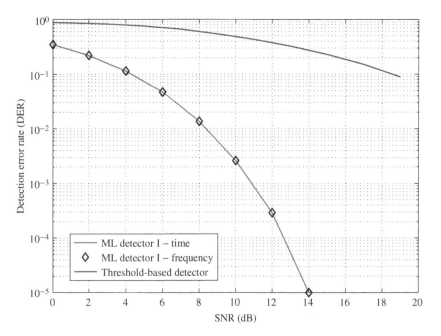

Figure 10.11 DER versus SNR without a guard-band between resonator frequencies.

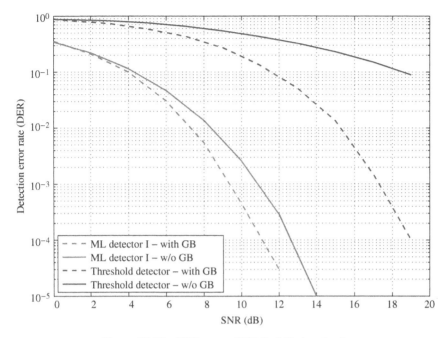

Figure 10.12 DER versus SNR for ML decoder 1.

the proposed method requires only 3 dB providing an SNR gain of 16 dB, which is equivalent to a tag reading range improvement by six times. However, in reality the maximum reading range is limited by the mismatches in the tag orientation, antenna gain, transmitted power, and receiver's dynamic range. At large distances, all these factors contribute to cause performance deterioration.

For further clarity, Figure 10.12 shows a comparison of the DER for both threshold method and ML decoder under the presence and absence of a guard-band (GB). At lower SNR levels (< 5 dB) such as noisy industrial environment, the proposed detection method on both resonator sets performs similarly. However, at higher SNR levels, resonators with a guard-band perform better for obvious reasons. The key observation is that the threshold-based detector has very poor performances at compact tag resonators (without guard-band); however, the proposed detector still has an accepted level of tag reading accuracy. Therefore, it can be argued that the proposed detection method allows the tag data bit capacity to be doubled without compromising the tag reading performance. In addition, the likelihood expression is valid for both time- and frequency-domain-based sampling.

10.6.2 System Model II

After obtaining satisfactory performances from *Model I*, the *System Model II* is tested with both the real (I) and imaginary (Q) components of the received

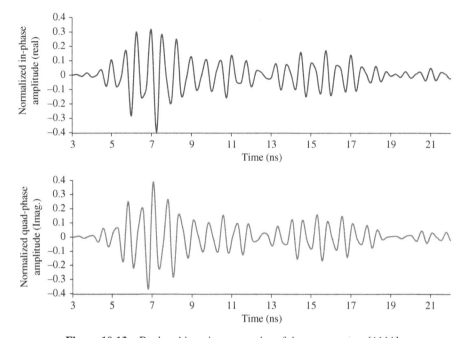

Figure 10.13   Real and imaginary samples of the tag response [1111].

signal. Figure 10.13 shows the real and imaginary components of the baseband signal obtained after I/Q demodulation for tag response of [1111] when a 60 MHz guard-band is used between resonator frequencies.

FFT is performed on these I/Q samples and the tag response in frequency domain is obtained. Figure 10.14 illustrates the magnitude and phase of the frequency-domain tag response. It can clearly be seen that the information is encoded in both the magnitude and phase. As a result, it can be concluded that unlike in *Model I*, *Model II* uses the information used in both the I/Q samples. Therefore, it is expected to outperform the detector derived for *Model I*. These time- as well as the frequency-domain samples obtained by applying fast Fourier transformation to the time samples are used to calculate the DER at different SNR levels. Figure 10.15 shows a comparison of the DER for both the threshold method and ML decoder 2 when a guard-band is presented. It can be seen that the threshold detector achieves a tag reading accuracy of 99.99% at an SNR of 17 dB. The same level of reading accuracy can be achieved at 6 dB using ML decoder 2. The ML detector provides an SNR gain of 11 dB at 99.99% accuracy level that can be related to the improvement of the reading range by a factor of 3.5.

On the other hand, at an SNR of 8 dB, the ML detector achieves and reading accuracy of 99.999% while the threshold detector manages only 70%. In addition, the proposed ML detector 2 has a tag reading accuracy of 95% at as low as 0 dB SNR. Therefore, it can also be concluded that the proposed ML detector 2 is performing well under low SNR levels. It is interesting to notice that ML detector 2 has an SNR

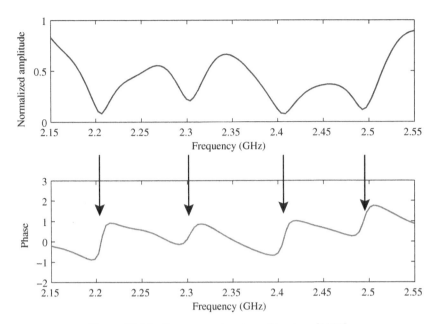

Figure 10.14 Frequency signature of tag type [1111].

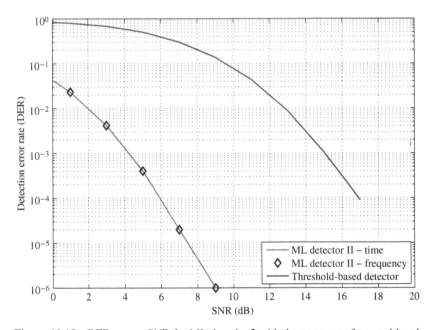

Figure 10.15 DER versus SNR for ML decoder 2 with the presence of a guard-band.

gain of 3 dB compared to ML detector 1. This is due to the fact that ML detector 2 uses in information available in both the real and imaginary components of the received signal, whereas ML detector uses only the real component of the signal. A detailed comparison of the results obtained in different models is presented later in this section.

Figure 10.16 shows a comparison of the DER for both threshold method and ML decoder 2 when a guard-band is removed. Similar to *System Model I*, the threshold-based detector performs very poorly when the guard-band is removed. At an SNR of 10 dB, threshold-based detector has a reading accuracy of 50% while ML detector 2 achieves well over 99.99% tag reading accuracy. On the other hand, it is obvious that ML detector 2 provides an enormous SNR improvement over the threshold detector. This significant improvement is mainly due to the fact that unlike threshold detector the likelihood expression derived in Equation (10.15) is the optimum decoder as it uses the information available in both the real and imaginary components of the signal. A comprehensive comparison of the performances of the proposed detection techniques is presented at the end of Section 10.6.

Figure 10.17 shows a comparison of the DER for both the threshold method and ML decoder 2 under the presence and absence of a guard-band. It is clear that for obvious reasons, both detectors perform well with the presence of a guard-band. However, once the guard-band is removed, threshold detector gets affected the most while ML decoder 2 still provides acceptable performances. ML detector performs well at lower noise levels regardless of the guard-band.

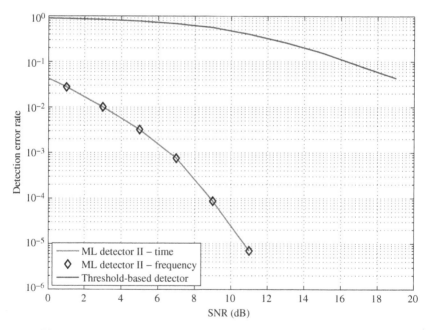

Figure 10.16   DER versus SNR for ML decoder 2 without a guard-band.

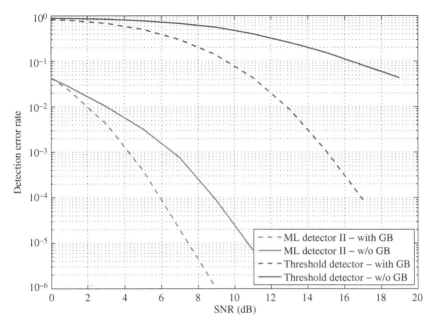

Figure 10.17   DER versus SNR for ML decoder 2.

In addition, it can be seen that even after removing the guard-band, *Model II* detector performs better than the existing threshold-based method. It can be interpreted as doubling the data capacity per unit bandwidth. In order to verify this claim, the simulations are repeated with eight resonators in the same bandwidth compared to four in the previous case. Figure 10.18 shows the calculated DER under 8 bits and compared against 4 bits tags with and without a guard-band. It can be seen that 4-bit tag with a guard-band has the best performance. Even though the 8-bit tag without a guard-band has the worst performance out of the three tags, it is comparable to that of the 4-bit tag without a guard-band. Hence, it is possible to conclude that the proposed detection algorithm doubles the tag data capacity.

10.6.3   System Model III

System Model III assumes an unknown channel with a known channel distribution. It is assumed that both the real and imaginary parts of the channel have the same Gaussian distribution due to symmetry. The mean and the standard deviation of the distribution are taken as 0.4 and 0.1, respectively. At different noise levels, DER is calculated based on both time and frequency samples and compared that with the baseline detection applied on frequency-domain samples.

Figure 10.19 shows a comparison of the DER for both threshold method and ML decoder 3 when a guard-band is presented. Unlike in previous two models, channel information is not available at the RFID reader. ML detector derived in this model uses the statistical properties of the channel for tag detection. Therefore, as expected,

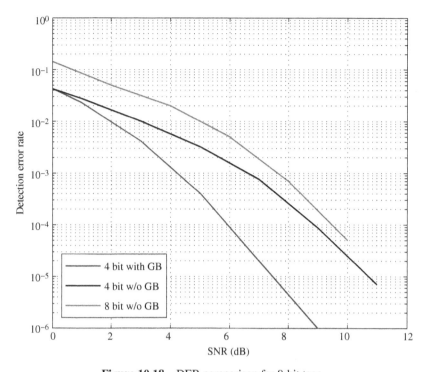

Figure 10.18   DER comparison for 8-bit tags.

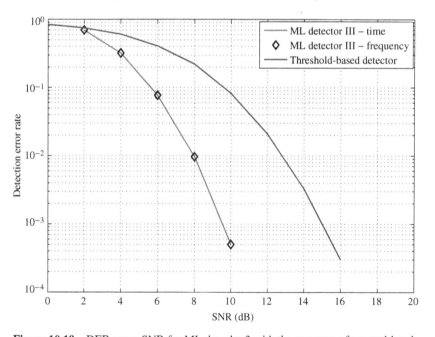

Figure 10.19   DER verus SNR for ML decoder 3 with the presence of a guard-band.

the performances are not very good as previous models, especially under low SNR scenarios. ML detector 3 achieves about 90% reading accuracy at SNR = 5 dB, which is still better than the 50% accuracy level of threshold-based detection. However, as SNR improves, the reading accuracy improves exponentially. For example, at an SNR of 10 dB, ML detector achieves an accuracy level of 99.95%, whereas threshold detector achieves only 90%.

Moreover, ML detector achieves an accuracy level of 99.9% at SNR = 9 dB while threshold detector needs 15 dB in order to achieve the same accuracy level. This is an SNR gain of 6 dB, which is equivalent to the improvement of the reading range by a factor of 2. In addition, similar to previous models, time- and frequency-based ML detectors provide the same results.

Figure 10.20 shows a comparison of the DER for both the threshold method and ML decoder 3 when there is no guard-band presented. As expected, threshold-based detector has high DERs. For example, the reading accuracy of threshold detector at SNR = 10 dB is about 50%, while the proposed ML detector achieves an accuracy level of 99% at the same SNR. Even though the tag reading accuracy at lower SNR is poor, it improves exponentially as SNR increases. On the other hand, the proposed ML detector has an SNR gain of 10 dB over threshold detector when both detectors are expected to achieve an accuracy level of 95%. This SNR gain provides an improvement in reading range by a factor of 3. It can be concluded that the proposed ML-based detector performs better than the threshold-based detector used in existing RFID readers.

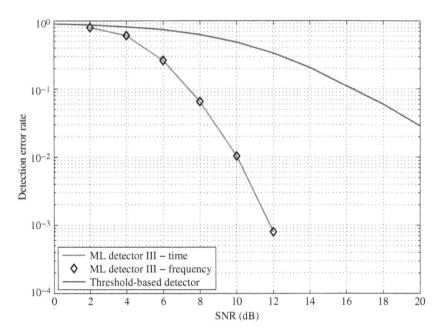

Figure 10.20 DER versus SNR for ML decoder 3 without a guard-band.

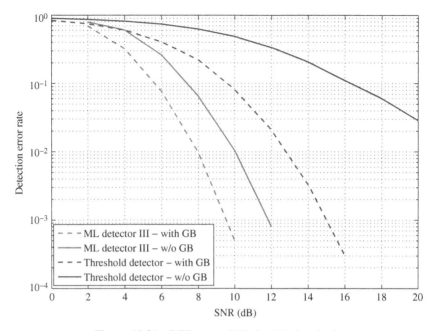

Figure 10.21   DER versus SNR for ML decoder 3.

Figure 10.21 shows a comparison of the DER for both threshold method and ML decoder 2 under the presence and absence of a guard-band.

It can clearly be seen that all of the detection methods failed to operate at lower SNR levels ($<$ 5 dB). So it is safe to conclude that the assumptions to represent both the real and imaginary parts of the channel in a Gaussian distribution *System Model III* is valid only for higher SNR levels ($>$ 5 dB). However, ML-based detection technique performs better as the SNR improves regardless of the guard-band. The guard-band provides on average an SNR improvement of 2 dB for obvious reasons. Finally, it can be concluded that the proposed ML detection method allows to double the bit capacity of the chipless RFID tags without compromising the reading accuracy.

10.6.4   System Model IV

As described earlier, no channel information is available at the reader in *System Model IV*. This ML-based decoder detects both the tag type and the channel simultaneously. Similar to the previous cases, DER is calculated based on both time and frequency samples and compared that with the baseline detection applied on frequency-domain samples.

Figure 10.22 shows a comparison of the DER for both threshold method and ML decoder 3 when a guard-band is presented. Unlike the *System Model III*, this model performs better even at lower SNR levels. In this model, no assumptions are made to represent the channel. Instead, channel values are estimated along with the tag

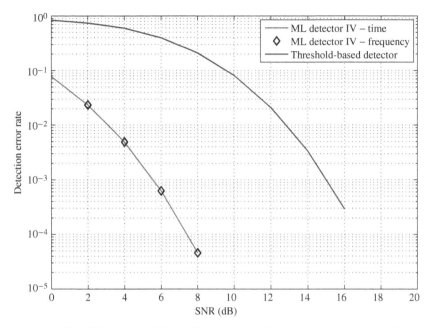

Figure 10.22   DER versus SNR for ML decoder 4 with the presence of a guard-band.

detection. It can be seen that, with the presence of a guard-band, ML detector achieves a reading accuracy level of 99.9% at SNR = 5 dB. However, threshold detector provides only 50% accuracy at the same SNR.

On the other hand, a reading accuracy of 99.9% is achieved by the proposed ML detector with an SNR gain of 10 dB over the threshold method. This results in improving the tag reading range by a factor of 3. In addition, both the time- and frequency-domain samples-based data provide the same performances as expected.

Figure 10.23 shows a comparison of the DER for both threshold method and ML decoder 3 when there is no guard-band presented. Following the trend in previous models, *System Model IV* performs well, even without a guard-band between resonance frequencies while the threshold detector has very poor performances. For example, at SNR =],5 dB, ML detector provides a reading accuracy of 99.5% while threshold detector provides on 20%. Apart from that, the proposed ML decoder provides an SNR gain of 17 dB over the threshold method when both are achieving a reading accuracy of 97%. This can be related to tag reading range being improved by a factor of 7. However, as explained earlier, this is possible only if the perfect tag orientation is achieved. The misorientation of the tag and the reader antennas may deteriorate the performance. After observing both aspects, it can be concluded that the proposed detection method allows to double the tag data bit capacity without compromising the tag reading performance.

Figure 10.24 shows a comparison of the DER for both threshold method and ML decoder 2 under the presence and absence of a guard-band.

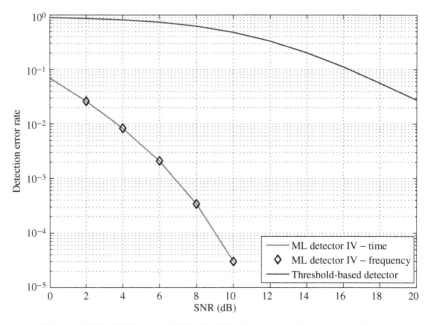

Figure 10.23   DER versus SNR for ML decoder 4 without a guard-band.

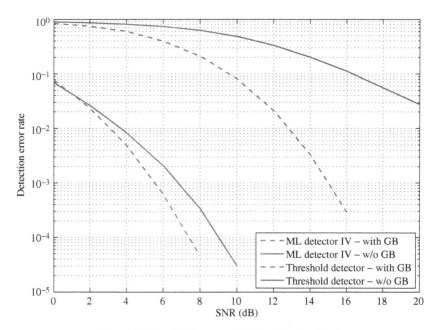

Figure 10.24   DER versus SNR for ML decoder 4.

A common feature of the results for *System Model IV* is that it performs well even under the lower SNR levels, regardless of the guard-band. As expected, at higher SNR levels, the guard-band provides an SNR gain of 2 dB. As mentioned at the beginning, ML detector 4 not only detects the tag but also estimates the channel.

10.6.4.1 Channel Estimation The channel estimation accuracy is analyzed next. In MATLAB, a random channel is generated and used for system simulation. Then ML decoder 4 is used to calculate an estimate of the channel and compared with the actual channel. Figure 10.25 compares the estimated channel values obtained under number of iterations with the actual channel realization when a guard-band is presented between resonators in the chipless tag. It can be seen that the complex channel estimations are centered around the actual channel realization and the accuracy level of the estimations are very high at SNR = 14 dB. However, it does not demonstrate a clear picture of the distribution of the channel estimate.

Figure 10.26 shows the probability distribution function (pdf) of the channel estimation for both real and imaginary components. It compares the estimated channel with the actual channel realization, which is given by white-colored asterisks (*).

It can be clearly seen that the mean of each distribution is the same as the actual channel realization. In addition, both distributions have very similar but low variances. Therefore, it can be concluded that the proposed ML detection method estimates the channel accurately when a guard-band is presented.

Figure 10.27 compares the estimated channel values with the actual channel realization when more tag bits are presented leaving no guard-band between resonator

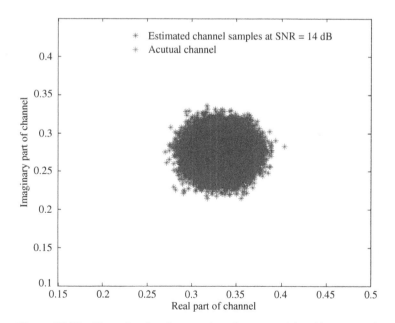

Figure 10.25 Channel estimation samples when a guard-band is presented.

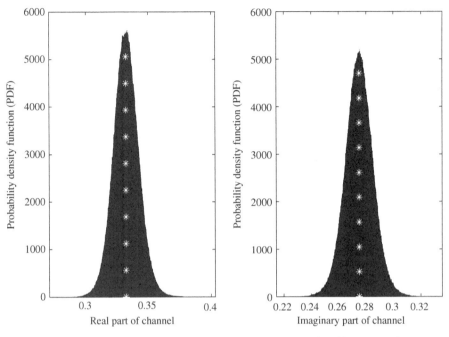

Figure 10.26   PDF of channel estimation when a guard-band is presented.

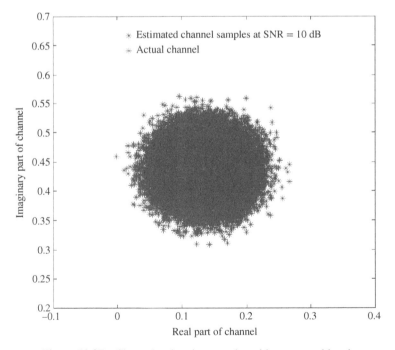

Figure 10.27   Channel estimation samples without a guard-band.

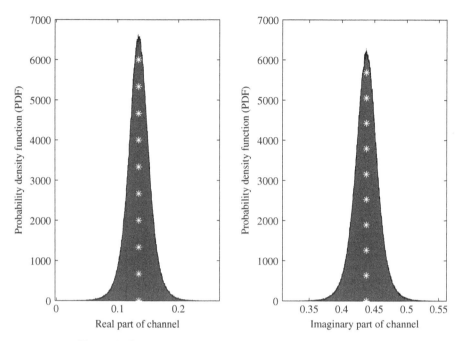

Figure 10.28 PDF of channel estimation without a guard-band.

frequencies in the chipless tag. The results show channel estimates obtained after a number of iterations at SNR = 10 dB. It is clear that even when the guard-band is not present, the proposed ML detector manages to estimate the channel accurately. The pdf of the estimates are calculated to closely observe the performance of channel estimation.

Figure 10.28 compares pdf of the channel estimation at SNR = 10 dB. Similar to the case with the guard-band, the mean of the distributions for both the real and imaginary components of the channel are approximately same as the actual channel realization. Moreover, the variances of both the real and imaginary components of the channel are very similar and low. The main difference between the cases having a guard-band and without is that the latter has a high variance compared to the former. It can be justified as the guard-band prevents interresonator interference causing less errors in the outcome of the ML detector used with tag responses having a guard-band.

10.6.4.2   Comparative Study   Finally, DER performances of all the detection techniques described are compared and discussed. Figure 10.29 compares the DER performances of all the detection when a guard-band is presented between the tag resonance frequencies. It can be seen that ML detector 2 has the best performance out of all the detectors presented. ML detector 2 assumes perfect channel knowledge is available at the reader and uses information available in both real and imaginary components of the signal. This is the optimum detector for the chipless RFID system and hence can be treated as the upper margin for tag reading accuracy performances. The

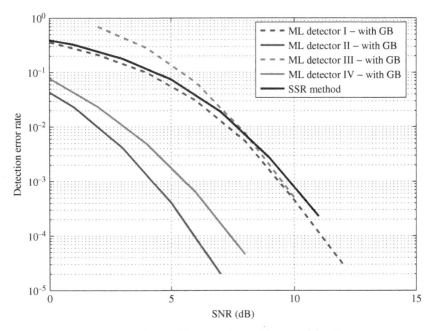

Figure 10.29   DER comparison with a guard-band.

next best detector is ML detector 4, where the channel is completely unknown to the reader. ML detector 4 is a powerful detector as it not only achieves a higher tag reading accuracy but also estimates the channel with a high accuracy. ML detector 1 is the next best performer, where it assumes perfect channel knowledge and uses only the information available in the real part. This decoder does not exist in reality as this is derived first as the fundamental decoder and every other decoder is an extension of it.

Signal space representation (SSR) method used in Ref. [13] has similar performances to ML detector 1. This can be explained as both detectors use the information encoded in only on the magnitude. However, only five most dominant basis functions are used for SSR and leaving out the rest causes the performances to be slightly degraded. It can be concluded that SSR method almost achieves its upper limit performances that are the performances of ML detector 1. Even though both SSR and the proposed ML-based detection techniques require the same number of computations, SSR has slightly lower computation complexity in each iteration. For example, ML detector 1 calculates the minimum distance using the total number of samples of the received signal while SSR does it using only five samples that are the coordinates in five-dimensional space.

Finally, ML detector 3 shows similar performances to ML detector 1 at higher SNR levels. Its performance at lower SNR is poor compared to other detection techniques. This pinpoints that the Gaussian distribution model used for each real and imaginary part of the channel is not very accurate at lower SNR levels.

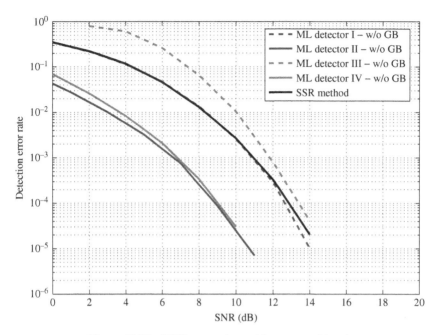

Figure 10.30   DER comparison without a guard-band.

Figure 10.30 compares the DER performances of all the detection techniques when there is no guard-band between the tag resonator frequencies. Similar to the case with a guard-band, ML detector 2 has the best performance. It is interesting to notice that ML detector 4 achieves almost the upper margin even though no channel information is available. SSR method for this scenario uses only the most significant seven base functions. It can be seen that with seven base functions, SSR method achieves its upper margin performances, which are the performances of ML detector 1. This confirms that the proposed detection methods perform better than the SSR method.

However, at lower SNR levels ($<$ 10 dB), ML detector 3 performs worse than ML detector 1. As the SNR increases ($>$ 10 dB), ML detector 3 performs similarly to ML detector 1. Similar to the case with a guard-band, even though *System Model* 3 models the system quite poorly at lower SNR levels, it is a valid detector for higher SNR levels. In addition, all ML-based detectors perform significantly better than the threshold-based detectors used in existing chipless RFID readers. Table 10.2 summarizes the results of all investigated methods so far.

10.6.5   System Model V

The likelihood detector derived under this model uses only the magnitude of the tag responses in the frequency domain. This tag detection technique is derived for an existing chipless RFID system that operates between 21 and 27 GHz. Figure 10.31 shows the tag response recorded by the chipless RFID reader.

TABLE 10.2   DER Comparison for Different Detection Methods

SNR		2	4	6	8	10	12
Threshold method	with GB	7.0E−1	6.0E−1	4.0E−1	2.0E−1	8.0E−2	2.0E−2
	w/o GB	9.0E−1	8.0E−1	7.0E−1	6.0E−1	5.0E−1	3.0E−1
SSR method	with GB	2.5E−1	1.0E−1	4.0E−2	6.0E−3	8.0E−4	N/A
	w/o GB	2.2E−1	1.2E−1	4.6E−2	1.3E−2	2.7E−3	3.3E−4
Model I	with GB	2.1E−1	1.0E−1	3.1E−2	5.5E−3	4.5E−4	3.0E−5
	w/o GB	2.2E−1	1.1E−1	4.7E−2	1.4E−2	2.6E−3	2.9E−4
Model II	with GB	1.0E−2	1.0E−3	1.0E−4	3.0E−6	N/A	N/A
	w/o GB	2.0E−2	6.0E−3	2.0E−3	3.0E−4	2.0E−5	N/A
Model III	with GB	7.0E−1	2.8E−1	6.6E−2	7.7E−3	5.0E−4	N/A
	w/o GB	8.0E−1	6.1E−1	2.6E−1	6.6E−2	1.0E−2	8.0E−4
Model IV	with GB	2.3E−2	4.9E−3	6.2E−4	4.6E−5	N/A	N/A
	w/o GB	2.6E−2	8.4E−3	2.1E−3	3.4E−4	3.0E−5	N/A

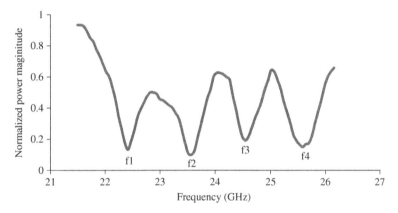

Figure 10.31   Magnitude of the tag response for tag [1111].

The tag response received in Figure 10.31 together with all possible 16 combinations are used for the detection algorithm derived for *System model V*. Table 10.3 shows the likelihood values obtained for each tag type. In order to avoid underflow, probabilities are calculated in log to the base 10.

It can be clearly seen from Table 10.3 that tag type [1111] has the highest likelihood out of the 16 combinations. Therefore, the proposed detection technique has managed to detect the tag bits successfully.

A comprehensive analysis is performed using MATLAB simulations and Figure 10.32 shows both the simulated and experimental DER variation at different SNR levels. In addition, the results are compared with the threshold-based detection technique. It can be seen that the proposed tag detection technique has superior

TABLE 10.3   Likelihood for each tag type.

Tag Type	Likelihood in Log Scale	Tag Type	Likelihood in Log Scale
[0000]	−35.41	[1000]	−28.54
[0001]	−28.50	[1001]	−19.77
[0010]	−27.76	[1010]	−20.70
[0011]	−22.87	[1011]	−15.59
[0100]	−29.76	[1100]	−24.22
[0101]	−22.47	[1101]	−16.87
[0110]	−22.28	[1110]	−15.58
[0111]	−17.31	[1111]	−11.06

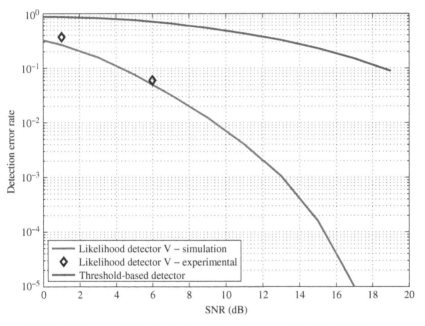

Figure 10.32   DER versus SNR for likelihood-based detector 5 for 21–27 GHz backscattering tag.

reading accuracy compared with the existing threshold-based detection technique. It is also important to notice that the experimental results agree with the simulated results. The experimental data is gathered only for two SNR levels as achieving low DERs at higher SNR levels are practically not feasible.

It can be concluded that the proposed tag detection technique for the magnitude-based chipless RFID reader achieves higher reading accuracy over the existing tag detection technique.

10.7 CONCLUSION

DER of a number of likelihood-based detectors were presented and compared against the threshold-based detector used in existing chipless RFID systems and the SSR method proposed in Ref. [13]. It is evident that all the likelihood-based detectors perform better than the threshold-based detector. ML detector 4 that jointly detects both the channel and the tag type achieves almost the optimum performance (ML detector 2). The improved performance of the proposed tag detection techniques directly relates to an increased tag reading accuracy at a given SNR level. On the other hand, it can also be represented as an increment in the reading range while achieving a particular goal of reading accuracy. Therefore, the improved performance can be represented either as increased reading accuracy or the reading range depending on the application.

However, there is a common drawback of all the likelihood-based detection methods discussed so far. All these methods require higher computation complexity compared to the primitive detection techniques such as threshold-based detection. For example, detecting a tag having N bits involves evaluating the likelihood expressions for 2^N number of occasions. Two computationally feasible tag detection techniques have been introduced in *Chapter 11* that can reduce the computation complexity from exponential to linear order without compromising on the chipless RFID tag reading accuracy.

REFERENCES

1. S. Preradovic, I. Balbin, N. Karmakar, and G. Swiegers, "Multiresonator-based chipless RFID system for low-cost item tracking," *IEEE Transactions on Microwave Theory and Techniques*, vol. 57, no. 5, pp. 1411–1419, May 2009.

2. S. Harma, V. Plessky, C. Hartmann, and W. Steichen, "PS-1 SAW RFID tag with reduced size," in *Ultrasonics Symposium, 2006. IEEE*, Vancouver, BC, Canada, Oct 2006, pp. 2389–2392.

3. Y.-Y. Chen, T.-T. Wu, and K.-T. Chang, "P5H-3 A COM analysis of SAW tags operating at harmonic frequencies," in *Ultrasonics Symposium, 2007. IEEE*, New York, NY, USA, Oct 2007, pp. 2347–2350.

4. S. Harma, V. Plessky, X. Li, and P. Hartogh, "Feasibility of ultra-wideband SAW RFID tags meeting FCC rules," *IEEE Transactions on Ultrasonics, Ferroelectrics, and Frequency Control*, vol. 56, no. 4, pp. 812–820, April 2009.

5. S. Preradovic and N. Karmakar, "Design of fully printable planar chipless RFID transponder with 35-bit data capacity," in *Microwave Conference, 2009. EuMC 2009. European*, Rome, Italy, Sept 2009, pp. 013–016.

6. IDTechEx, *Chip-less RFID? The end game*, 2006 (accessed December 24, 2014). [Online]. Available: http://www.idtechex.com/products/en/articles/00000435.asp.

7. K. C. Jones, *Invisible tattoo ink for chipless RFID safe, company says*, 2007 (accessed December 24, 2014). [Online]. Available: http://www.electronics-eetimes.com/en/invisible-tattoo-ink-for-chipless-rfid-safe-company-says.html?cmp_id=7& news_id= 196900063.

8. S. Innovations, *RFID Tattoo: Somark of the Beast*, 2008 (accessed December 24, 2014). [Online]. Available: http://www.rap-con.com/signs/rfid-tattoo-somark-beast.

9. I. Jalaly and I. Robertson, "RF barcodes using multiple frequency bands," in *Microwave Symposium Digest, 2005 IEEE MTT-S International*, Long Beach, CA, USA, June 2005, pp. 139–142.

10. J. McVay, A. Hoorfar, and N. Engheta, "Space-filling curve RFID tags," in *Radio and Wireless Symposium, 2006 IEEE*, San Diego, CA, USA, Jan 2006, pp. 199–202.

11. S. Preradovic, I. Balbin, S. M. Roy, N. C. Karmakar, and G. Swiegers, "Radio frequency transponder," Australian Provisional Patent Application P30 228AUPI, April, 2008.

12. G. Janssen and R. Prasad, "Propagation measurements in an indoor radio environment at 2.4 GHz, 4.75 GHz and 11.5 GHz," in *Vehicular Technology Conference, 1992, IEEE 42nd*, May 1992, pp. 617–620 vol. 2.

13. P. Kalansuriya, N. Karmakar, and E. Viterbo, "Signal space representation of chipless RFID tag frequency signatures," in *Global Telecommunications Conference (GLOBE-COM 2011), 2011 IEEE*, Houston, TX, USA, Dec 2011, pp. 1–5.

11

COMPUTATIONALLY FEASIBLE TAG DETECTION TECHNIQUES

11.1 INTRODUCTION

Maximum likelihood (ML)-based detectors generally produce better tag detection performances as they utilize all information available, prior to making a decision. However, one of the main drawbacks of ML detectors is its higher computation complexity. In the proposed tag detection techniques, they compare the received signal with all possible tag combinations and select the one with the highest probability as the detected tag. If a radio frequency identification (RFID) tag has N bits, they compare the received signal with all 2^N tag combinations to calculate individual probabilities. In the case of tags being used to identify the object category not the object itself, the number of bits required in a tag can be small. In such cases, direct application of the tag detection techniques presented in *Chapter 10* may be feasible. However, when the number of bits in the tag is large, computation complexity is increasing exponentially, hence utilizing the tag detection techniques presented in *Chapter 10* may not be feasible. Higher computation complexity brings up two main challenges. First, the RFID reader needs higher computation capability to evaluate the likelihood expressions derived in *Chapter 10*. The second challenge is the increased computation time, which directly affects the tag reading time.

Regarding the former challenge, it can afford to deploy higher computation complexity at the RFID reader, with ever-evolving low-cost single-board computers available in few tens of dollars. However, some of the repeated calculations can be preprocessed and stored in a lookup table for real-time use. In addition, memory

Advanced Chipless RFID: MIMO-Based Imaging at 60 GHz–ML Detection, First Edition.
Nemai Chandra Karmakar, Mohammad Zomorrodi, and Chamath Divarathne.
© 2016 John Wiley & Sons, Inc. Published 2016 by John Wiley & Sons, Inc.

required to store 2^N tag responses is manageable with few megabytes. For example, if each tag response has 100 samples and each sample is denoted by a signed short number format (16 bits), then the memory required for each tag response can be calculated as 200 bytes. For 10-bit tags, there are 1024 tag combinations. Therefore, the memory required for storing all possible combinations for 10-bit tags is only 200 KB. Hence, it can be argued that memory and the processing power can be overcome with the readily available hardware.

However, in practical applications, 10-bit tags are not sufficient to tag individual items, for example, grocery items in a superstore. In such applications, large data bits in the order of 30–60 bits are needed to tag each item with its serial number. Therefore, an efficient tag detection algorithm is to be developed for practical applications.

Computation time is mainly dominated by 2^N number of calculations performed for each tag combination. This chapter describes two techniques that are capable of reducing the total number of computations. The first technique reads bit by bit in the tag rather than reading all bits at the same time. The second technique uses a trellis tree-based Viterbi decoding method to reduce the number of computations from 2^N to $8 \times N$.

The rest of the chapter is organized as follows. First, the proposed bit-by-bit tag detection technique is presented. Next, the trellis-based Viterbi decoding method is explained. Then the simulation setup is described and the results for each method are presented in the following section. Finally, the chapter is concluded and recommendations for single input single output (SISO) chipless RFID tag reading are presented.

11.2 BIT-BY-BIT DETECTION METHOD

The tag detection methods presented in *Chapter 10* first evaluate the likelihood expression for all the possible tag combinations and the one with the highest likelihood is selected as the detected tag. It has been proved in *Chapter 10* that the proposed exhaustive detection methods work based on both time- and frequency-domain samples. Frequency spectrum of the multiresonator tags has a unique advantage when it comes to observing the individual resonator's contribution to the overall tag response. Unlike in time-domain tag responses, frequency-domain tag responses have dedicated blocks of samples representing contribution from each resonator. It is more evident when there is a guard band to minimize any interresonator interference (IRI). The guard band helps to make a decision only based on the frequency samples in the given block. Therefore, samples in each frequency block are observed separately and the presence or the absence of the resonators in that block is detected using the highest likelihood. The process is described in detail next.

The first step is to separate the frequency-domain samples into N blocks representing the N number of tag data bits. Each block ($\bar{y}_k$) has only two possibilities, namely the response when a resonator is present (bit "1") and a resonator is absent (bit "0"). The kth resonator response in the frequency band around the resonance frequency is given by $\bar{S}_{k,1}$. The absence of resonance is denoted by a signal ($\bar{S}_{k,0}$) having a constant amplitude of 1 and a linear phase variation. For example, a 4-bit tag

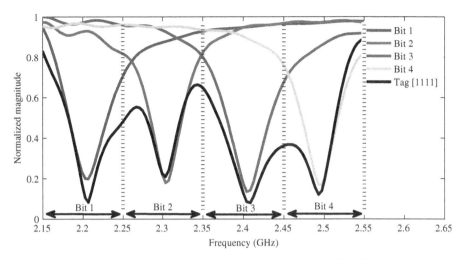

Figure 11.1   Bit-by-bit detection for a tag having [1111].

would have at most four resonator dips. In each block, the tag response is compared against the response when a resonator is present and absent. Figure 11.1 shows the comparison of overall tag response with each resonator response in individual blocks. The comparison is shown only in the magnitude; however, it can be extended to the phase as well. Phase comparison has to be performed with care. As there could be a phase offset in the overall tag response compared to individual resonator response, it should be removed before performing the real and imaginary comparisons.

Then the likelihood of receiving a bit "1" and a bit "0" is calculated using the likelihood function derived in *Chapter 10* and the decision is made based on the highest probability. It can be represented as follows:

$$\max_{\bar{S}_i = \{\bar{S}_{k,1}, \bar{S}_{k,0}\}} \Pr(\bar{y}_k | \bar{S}_{k,i}) \tag{11.1}$$

This process is repeated for all N blocks until all the tag bits are detected. A flowchart of the bit-by-bit tag detection technique is presented in Figure 11.2. The likelihood functions of all five frequency domain-based detection techniques derived in *Chapter 10* are compatible with the proposed bit-by-bit detection method assuming there is a guard band between the resonator frequencies.

Since the tag bits are detected sequentially, this method can be termed as *serial bit reading*, whereas the methods presented in *Chapter 10* read all tag bits simultaneously hence termed as *parallel bit reading*. The advantage of serial bit reading is less computation complexity compared to *parallel bit reading*. In this method, detection of each tag data bit requires evaluating the likelihood function two times: with the presence and the absence of the resonator. If N is the total number of tag bits, then the number of likelihood function evaluations required in the serial reading is $2 \times N$, whereas parallel reading requires 2^N iterations. It can be seen that the order

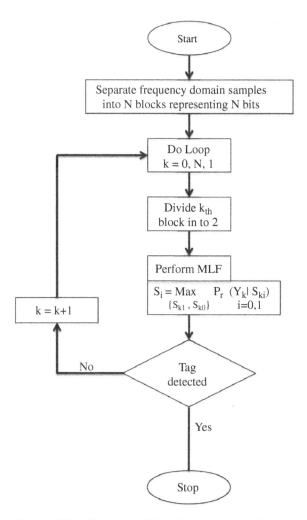

Figure 11.2   Flowchart of bit-by-bit detection technique.

of complexity in serial reading is linear, which is a significant improvement over the exponential complexity presented in the exhaustive ML-based detection techniques presented in *Chapter 10*. Also only the samples in a given frequency block are evaluated, the length of the vectors used in the calculations is smaller, hence less computation capabilities are required. In addition, the memory requirement to store the tag responses has reduced from exponential (2^N) to linear ($2 \times N$). As a result, the *serial bit reading* method also benefits from having less memory management in the RFID reader.

However, one disadvantage of the *serial bit reading* method proposed is that this method assumes a guard band to minimize IRI. That limits the bit capacity available for the tags. Moreover, in this method, only the information embedded in magnitude is used for decision making, hence it is a suboptimal detector. With the presence of the guard band, most of the data are confined in each frequency block itself. However, there could be more data available in the neighboring frequency blocks in terms of interference. This method does not utilize those extra information for decision making hence not an optimal detector and as a result this method cannot be treated as an ML method. These shortcomings are addressed in the trellis-tree-based Viterbi decoding detection presented in the following section.

11.3   TRELLIS-TREE-BASED VITERBI DECODING

11.3.1   Introduction

The bit-by-bit tag detection method presented in Section 11.2 assumes there is minimal or no IRI. In other words, the presence of a guard band minimizes the interference from the neighboring resonators. Therefore, the above-mentioned technique is valid only when there are guard bands presented in multiresonator tag design. As discussed earlier, this will limit the bit capacity for a given bandwidth. Moreover, the detector presented in Section 11.2 is suboptimal as some useful information is thrown out by not considering the neighboring frequency blocks. Trellis-tree-based Viterbi decoding technique discussed in this section can be treated as an optimal decoder because of its utilization of the most useful information available in the tag. Trellis tree creates a specific orthogonal code structure that eliminates redundant calculations and achieves exhaustive likelihood performances using limited calculations.

Viterbi decoding [1] is a forward error correction decoding method. A transmitter sends extra information along with data for better error performance. It assumes the encoding scheme involves convolutional coding. In traditional Viterbi decoding, a convolutional encoding system is selected and that limits the number of allowed state transitions and their outputs. The basic idea here is to reduce the computational complexity by disregarding the invalid state transitions. The number of output bits generated per input bit is determined by the code rate used for encoding. For example, $\frac{1}{2}$ rate involves doubling the number of output bits compared to the input bits. However, in chipless RFID systems, data bit capacity is relatively small and cannot afford to have lower code error rates. Therefore, the traditional Viterbi decoding cannot be applied directly. A novel approach needs to be used to efficiently utilize the Viterbi decoding algorithm. In the following section, the proposed approach is explained.

11.3.2   Signal Model

As discussed earlier, multiresonator-based chipless RFID tags consist of a number of resonators, which are designed to resonate at unique frequencies in the given bandwidth. The presence or absence of a resonator is used to encode data bit "1" or "0," respectively. For minimum interference with the neighboring resonators, a guard band is used between two resonance frequencies. This will limit the number of resonances placed in the given bandwidth hence limiting the total bit capacity in the tag. However, if the resonator frequencies in the chipless RFID tag are placed close to each other without a guard band, the individual resonances interfere with each other. This is very similar to intersymbol interference (ISI) in communication channels [2]. The proposed Viterbi algorithm is designed to perform under IRI, which means no guard band is required when designing the tags. This will allow improving tag data bit capacity.

In frequency domain, the final tag response can be treated as the product of individual resonator responses. If the final tag response has N number of samples and L number of resonators, each resonator band has on average $\frac{N}{L}$ number of samples.

It can be observed that when the guard band is removed, the ith resonator response $(\bar{X}_i)$ is mainly interfered by the neighboring two resonators. The interference from the resonators farther away can be neglected. In physical design approach, resonators with neighboring resonance frequencies are placed apart to reduce interresonator mutual coupling. Therefore, the actual response at the ith resonator response in frequency domain can be interpreted as a product of three resonators as shown in Equation (11.2). The assumption here is that the individual resonator responses are treated as *orthogonal functions* and the multiple resonator responses can be calculated using the product of orthogonal individual resonator responses. As a result, this method uses almost all the useful information available unlike in the bit-by-bit detection method discussed in Section 11.2. Therefore, the ith resonator response $(\bar{S}^{(i)})$ can be represented in Equation (11.2).

$$\bar{S}^{(i)} = \bar{X}_{i-1} \times \bar{X}_i \times \bar{X}_{i+1} \tag{11.2}$$

For simplicity, *Signal Model I* with frequency-domain samples is used to explain the operation principle of trellis-tree-based Viterbi decoding method. Therefore, the received signal $\bar{y}$ is given by

$$\bar{y} = h\bar{S}_m + \bar{\omega}$$

Notations are as the same as in *Chapter 10*. Likelihood function for *Signal Model I* is similar to the one derived in Equation (10.1) with the exception of using only a portion of the overall tag response for evaluation. As mentioned earlier, in frequency domain, each resonator has a bandwidth given by $\frac{N}{L}$ number of samples. Then likelihood function is evaluated using $3 \times \frac{N}{L}$ number of samples. The modified received signal is given by $\bar{y}_b$. $\bar{S}_b$ is one of the $8(2^3)$ possible combinations for the three resonators considered. The modified likelihood function can be written as in Equation (11.3) and it is optimized over the eight possible combinations of $\bar{S}_b$.

$$\max_{\bar{S}_b} \Pr(\bar{y}_b | \bar{S}_b) = \min_{\bar{S}_b} \left((\bar{y}_b - h\bar{S}_b)(\bar{y}_b - h\bar{S}_b)^T \right) \tag{11.3}$$

Now that the likelihood for a given condition can be performed, the operation of the trellis-tree-based Viterbi decoding algorithm is described next.

Figure 11.3 illustrates both states and the transitions allowed in a trellis tree. Most significant bit (MSB) of the tag data bits is considered as the first bit and the least significant bit (LSB) as the last bit. A state is constructed from two neighboring bits in

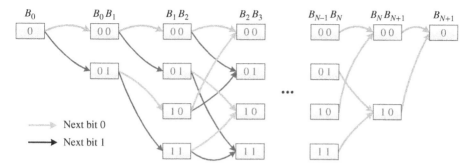

Figure 11.3 Operation of Trellis-tree-based Viterbi detection technique.

the tag. The first bit in the state is always closer to the MSB than the second. Each bit is repeated in the neighboring states as shown in Figure 11.3. In each bit position, there is a maximum of four possible states, hence four rows in the diagram. Each column denotes the bit position in a tag. Therefore, the third column represents all possible states considering the first (MSB) and second bits in the tag. It can be seen that at the two ends of the tree only a limited number of states are valid. It is explained using the initial and final conditions. As explained earlier, each resonator bit is interfered by the neighboring bits. The first resonator does not have any neighboring resonators to the left, hence the initial condition (B_0) for the analysis is assumed to be having a bit "0." Similarly, the final condition (B_{N+1}) is also assumed to be having a bit "0."

Although there are many states in the trellis tree, only a limited number of transitions are possible. Figure 11.3 shows state transition from the left to right only, even though both directions are possible. Transition between two states involves three bits. For example, transition between two states in $[B_1 B_2]$ and $[B_2 B_3]$ associates first 3 bits of the tag (B_1, B_2, and B_3). Transition between each state depends on the next bit value. For example, transition from state $[0\ 0]$ to state $[0\ 1]$ requires the next bit to be bit "1." No transition is possible from $[0\ 0]$ state to $[1\ 0]$ state as the first bit of the next state must have current bit, which is 0. It can be easily seen that a maximum of eight transitions are allowed between any two columns. The state transitions are further restricted at the two ends of the tree with "0" initial and final conditions.

Each state transition is associated with a likelihood given by Equation (11.3). For example, transition from $[0\ 0]$ state in the third column to $[0\ 1]$ state in the fourth column assumes $B_1 = 0$, $B_2 = 0$, and $B_3 = 1$. The likelihood for that transition can be calculated using $3 \times \frac{N}{L}$ long vectors of $\bar{y}_b$ and $\bar{S}_b$, which is the frequency response obtained with only the first three resonators presented in the tag.

The algorithm starts from state $[B_0]$ and likelihood for each state transition is calculated first. Until state $[B_1 B_2]$, each state has only one allowed transition. Therefore, there is only one path from the initial state to each of the states in $[B_1 B_2]$. The total likelihood of the path can be calculated by taking the product of each transition likelihood along the path. Likelihood of transiting via this path is stored at each state and named as the *state probability* from now onward. In addition to the state probability, states also store the previous state of the path. However, any state in $[B_2 B_3]$ and onward can be reached using two different states of the previous column. Then the path from the initial state to a state in $[B_2 B_3]$ has two options. Likelihood for each path is calculated and the one with the highest likelihood is stored as the *state probability*. Since previous states already have the best possible path to the initial state, previous *state probabilities* and the transition likelihood can be used for current *state probability* calculation. Each of the path probabilities for current state is given by the product between the previous *state probability* and the matching transition likelihood. Then, the highest state probability among the two paths is stored in the state together with the respective previous state.

Once all the *state probabilities* are calculated and the final state is reached, the decoding algorithm works from the final state toward the initial state as shown in Figure 11.4. The final state has stored its state probability and the previous state $(B_N B_{N+1} = [1\ 0])$ with the highest path likelihood. The common bit between the previous state $(B_N B_{N+1} = [1\ 0])$ and the final state $(B_{N+1} = \text{"0"})$ is the decoded bit in the optimum path. Therefore, B_{N+1}, which is the final condition, is decoded correctly as "0." Similarly, if the previous state stored in $(B_N B_{N+1} = [1\ 0])$ is $(B_{N-1} B_N = [1\ 1])$ then the common bit between the two states is $(B_N = \text{"1"})$ and the LSB of the tag can be decoded as "1." The process of traversing the most likelihood path continues until the initial state is reached as shown in Figure 11.4. Along the way tag bits are decoded from LSB to MSB. The trellis-tree-based Viterbi decoding process is depicted in Figure 11.5.

At the beginning, we assumed that once the guard band is removed, a resonator is interfered only by two neighboring resonators. Therefore, all information required to decode a certain bit is only available in the current and its two neighboring resonators. Since trellis-tree-based Viterbi decoding is using these three resonators to make decision, it can be treated as an optimal (ML) detection technique with a less computation complexity. Reduction in the computation complexity is achieved by disregarding the invalid transitions. The total number of likelihood function evaluations (valid transitions) is $8N - 4$. It can be seen that the proposed detection method is able to reduce the computation complexity from exponential order to a linear order without compromising the tag bit capacity.

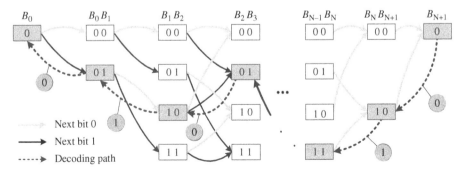

Figure 11.4   Viterbi decoding in a trellis tree.

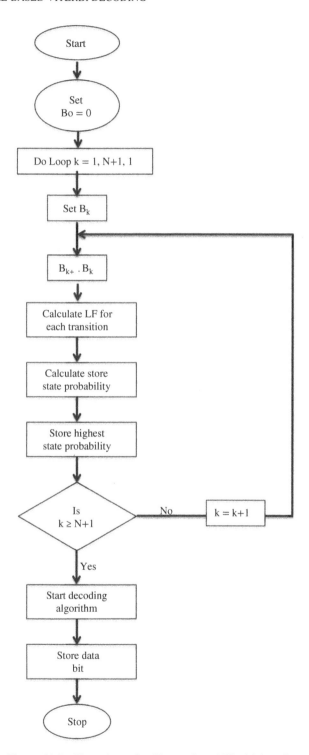

Figure 11.5 Flow chart of trellis-tree-based Viterbi decoding.

Then a simulation setup is designed using CST and MATLAB to verify the proposed tag detection techniques. The simulation setup is described next.

11.4   SIMULATION SETUP

The validity of the aforementioned tag detection techniques is verified using CST and MATLAB simulations.

11.4.1   CST Simulation

The steps carried out in the simulation are given in Figure 11.6. First, an interrogating signal is generated to provide a flat frequency response in 2.2–2.6 GHz frequency range. Four resonators are designed using CST with resonating frequencies as shown in Table 11.1. Then the combinations of resonators are placed besides a microstrip line to cover all possible tag IDs. One end of the microstrip line is fed with the interrogating signal and the tag responses are collected at the other end. These collected tag responses are saved in a lookup table for the algorithms to be used later. More details about the tag design were presented in *Chapter 9*.

11.4.2   MATLAB Simulation

Then the tag responses are fed through a channel that is given by the product of the forward and reverse channels of the RFID system. In order to prove the concept, only the detection technique derived for *System Model II* is used, which utilizes information

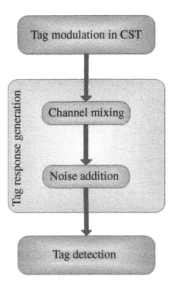

Figure 11.6   Flowchart of the MATLAB simulation.

TABLE 11.1 Simulation Parameters

Parameter	Value
Center frequency	2.4 GHz
Total bits encoded in a tag	4 bits
Flat frequency response	400 MHz
Bandstop filter attenuation	10 dB
Bandstop filter 3 dB bandwidth	50 MHz
Guard band	50 MHz
Resonance frequency set 1 (MSB to LSB)	[2.2, 2.3, 2.4, 2.5] GHz
Resonance frequency set 2 (MSB to LSB)	[2.34, 2.38, 2.42, 2.46] GHz
Channel mean	0.4
Channel standard deviation	0.1
Number of iterations	Up to 10,000,000

embedded in both amplitude and phase. The same principle can be extended to other models as well. The channel realization for *System Model II* is taken from a normalized distribution with a certain mean and variance. Finally, Gaussian noise is added to the resultant signal according to the specified signal-to-noise ratio (SNR). Simulation parameters are given in Table 11.1.

When calculating SNR, the signal power is taken as the average power of all the tag combinations. It can be summarized as follows:

$I(t)$ — interrogating signal
h_f — forward channel
h_r — reverse channel
h — product of forward and reverse channels
$F_m(t)$ — impulse response of the mth filter
$S_m(t)$ — m^{th} filter response
$y(t)$ — received signal
$\omega(t)$ — noise added at the reader.

Then the received signal can be represented using Equation (11.4).

$$y(t) = \left[[h_f I(t)] * F_m(t) \right] \times h_r + \omega(t)$$
$$= h_f h_r \times [I(t) * F_m(t)] + \omega(t)$$
$$= h S_m(t) + w(t) \qquad (11.4)$$

The power of each tag response is calculated and averaged to obtain the average power of a given tag response. For example, 2-bit tags have four different tag responses ($S_m(t)$) and power of each tag response is calculated and averaged to obtain the average power of 2-bit tag responses. Then the average tag response power is multiplied using the channel to calculate the average signal power available at the reader. For a given SNR, noise power is calculated using this available signal power.

MATLAB simulation parameters are outlined in Table 11.1. Four bandstop filters are used, and MSB corresponds to the lowest resonance frequency and LSB to the highest.

I & Q demodulation is performed with the received signal at the RFID reader. Then the two output time-domain signal vectors are used to evaluate likelihood expressions for each detection technique. In order to verify their frequency-domain performances the time-domain vectors are converted using fast Fourier transform (FFT). Then these frequency-domain samples are used for tag detection. Finally, the DER is calculated for each tag detection technique at different SNR levels. Existing chipless RFID systems use a threshold-based detection technique based on frequency-domain-based samples. The presence and absence of each resonator in this method is detected based on a magnitude threshold at corresponding resonator frequency bands. In the MATLAB simulation, this threshold-based detection technique is implemented and the DER is calculated to compare the performances of the proposed detection techniques. In addition, the average computation time for one reading using the above-mentioned MATLAB simulation setup is also calculated.

11.5 RESULTS

The simulations are carried out as given above and the results are presented in the following two sections. The first section presents the detection error rate performances under different noise levels. The second section compares the computation time for each detection method.

11.5.1 Detection Error Rate (DER)

System Model II in Section 10.2.2 utilizes all the information available when making a decision. Therefore, *System Model 2* is used for verification in this section. The expression derived in Equation (10.15) is used to evaluate the likelihood of a transition from one state to the other in trellis tree diagram.

Figure 11.7 compares the DER results for System model 2 under five cases. First, it presents the DER using exhaustive ML detection method derived in Chapter 10 with the presence and absence of a guard band. Then the trellis tree diagram having less computation complexity is used to calculate DER with and without a guard band. Finally, the above four cases are compared with the bit-by-bit detection method when a guard band is presented.

It can be seen that trellis decoding method has very similar results to the fully optimal detection method, that is, ML detector 2. As a result, it can be concluded that the trellis decoder is a fully optimal detector. As expected bit-by-bit detector has the poorest performance.

11.5.2 Computation Time

The main drawback of exhaustive likelihood-based detection techniques is its exponential computation complexity as the number of tag bits increases. Figure 11.8

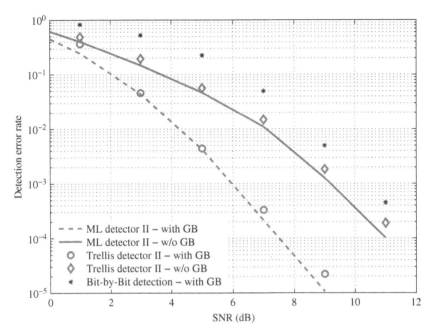

Figure 11.7 DER comparison for 10-bit tags.

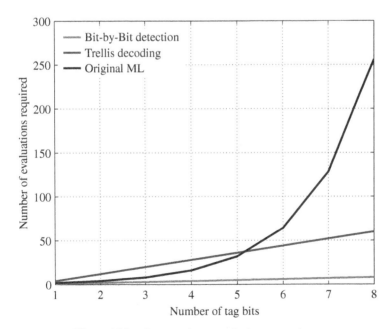

Figure 11.8 Computation complexity comparison.

presents the number of computations required for each tag detection type against the number of tag data bits. It can be noticed that both the trellis decoding and bit-by-bit detection technique has linear computation complexity against the exponential computation complexity of original likelihood detection methods. As expected, bit-by-bit detection technique has the lowest computation complexity, whereas trellis decoding provides a manageable complexity at higher tag bits.

Based on the DER and computation complexity results presented earlier, it can be concluded that, bit-by-bit detection method is preferred under very high-SNR environments, with larger tag bits having a guard band. In low-tag data capacity applications, original ML detection techniques are preferred. trellis decoding technique is the favorable choice when the tag data capacity is high.

11.6 CONCLUSIONS

Computationally feasible two tag detection techniques have been proposed to reduce the computation complexity from exponential to linear in *Chapter 11*. It is found that the bit-by-bit detection method, which is a suboptimal detection method, performs successfully when a resonator guard band is used in tag design. It is shown that the computation time has significantly dropped compared to exhaustive ML detection methods without compromising the reading accuracy. In addition, a fully optimal trellis-tree-based Viterbi decoding technique has been introduced to reduce the computation complexity from exponential to linear order while achieving a similar reading accuracy to original likelihood detection techniques.

REFERENCES

1. A. Viterbi, "Orthogonal tree codes for communication in the presence of white Gaussian noise," *IEEE Transactions on Communication Technology*, vol. 15, no. 2, pp. 238–242, April 1967.
2. R. Scolaro, "Error probabilities for orthogonal multiplex systems in the presence of inter-symbol interference and Gaussian noise," *IEEE Transactions on Communications*, vol. 25, no. 5, pp. 549–557, May 1977.

12

SIGNAL PROCESSING FOR MIMO-BASED CHIPLESS RFID SYSTEMS

12.1 INTRODUCTION

In *Chapter 8*, several chipless tags have been studied and it is concluded that multiresonator-based chipless tags will be further investigated in order to incorporate them with the proposed MIMO-based chipless RFID system. The main reason is that the multiresonator tags are reported as one of the chipless tag types with the largest data capacity. Also, Monash Microwave, Antennas, RFID and Sensors Laboratory (MMARS) has all the required equipment and facilities for tag fabrication and testing.

An overview of the proposed MIMO-based chipless RFID system is illustrated in Figure 12.1. The chipless tag used in the proposed system is multiresonator based, where each resonator acts as a bandstop filter introducing a frequency signature to the tag.

The tag considered earlier has one receiving antenna (Rx) and two transmitting antennas (Tx_1 and Tx_2) cross-polarized with Rx as shown in Figure 12.1. The RFID reader has one transmitting antenna (Tx) that is cross-polarized with its two receiving antennas (Rx_1 and Rx_2), hence minimizing the coupling between transmitting and receiving antennas. The transmitting antenna of the reader (Tx) and the receiving antenna of the tag (Rx) are copolarized that are already cross-polarized with the transmitting antennas of the tags (Tx_1 and Tx_2) so that, the undesired coupling throughout the system is minimized. As a result, when the reader transmits, it is safe to assume that only Rx of the tag receives the signal, hence forming a single input single output (SISO) channel from the reader to the tag named as the forward channel hereafter.

Advanced Chipless RFID: MIMO-Based Imaging at 60 GHz–ML Detection, First Edition.
Nemai Chandra Karmakar, Mohammad Zomorrodi, and Chamath Divarathne.
© 2016 John Wiley & Sons, Inc. Published 2016 by John Wiley & Sons, Inc.

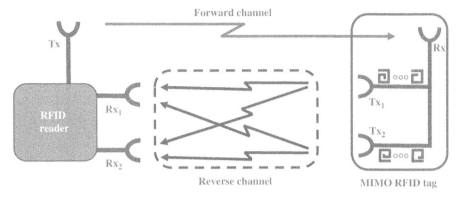

Figure 12.1   MIMO-based chipless RFID system.

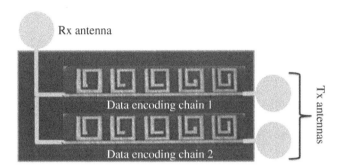

Figure 12.2   MIMO tag.

Then the received signal at the tag will be divided into two parts using an equal power divider. Each RF component will then travel toward its transmitting antenna surrounding the frequency resonators as shown in 12.2. When an RF signal travels surrounding the resonators, the resonators start to resonate at their resonating frequency, hence losing the power in the corresponding frequency of the signal. Once the RF signal reaches the end of the transmission line, it contains the frequency signature of the tag and this process is called tag modulation hereafter. The tag-modulated RF signals will then be transmitted back to the reader by using each transmitting antenna of the tag. There will be two different signals transmitting from the tag toward the reader and the reader will receive them using two receiving antennas (Rx_1 and Rx_2) having the same polarization to that of the transmitting antennas of the tag. This forms a 2×2 multiple input multiple output (MIMO) channel and is called reverse channel hereafter.

Tag detection involves two stages. First stage is about MIMO decomposing. When the RFID reader receives two streams of signals from its two receiving antennas, those signals are already mixed with the 2×2 MIMO channel. First MIMO decoding techniques are used to decompose these already mixed two signal streams. Once they are separated, in second stage, ML-based tag detection techniques are used to identify encoded tag data.

The resonator combination used in each branch of the MIMO tag can perform tag modulation individually. As a result, the bit capacity can be improved by several factors if the tag contains multiple branches. For example, a MIMO tag with two branches can double the tag bit capacity using the same frequency band of the SISO tag. This becomes achievable, thanks to the MIMO decomposing algorithms, which will be discussed in the following section.

12.2   MIMO DECOMPOSING TECHNIQUES

The signal model used in the proposed MIMO-based chipless RFID system is shown in Figure 12.3 and explained in this section.

If the interrogating signal is denoted by $x(t)$ and the forward channel by h_f, then the received signal at the tag $r_t(t)$ can be represented using (12.1), whereas $n_t(t)$ is the noise added by the receiving antenna at the tag.

$$r_t(t) = h_f x(t) + n_t(t) \tag{12.1}$$

Then the received noisy signal is divided into two power-equal components tx'_1 and tx'_2.

$$tx'_1 = tx'_2 = \frac{1}{\sqrt{2}} r_t(t)$$

These two identical signals are tag modulated using independently selected resonator combination. Assume that the equivalent bandstop filters of the resonator combinations in each branch are given by $f_1(t)$ and $f_2(t)$. Then the final tag response in each branch $(tx_1(t)$ and $tx_2(t))$ can be calculated as (12.2).

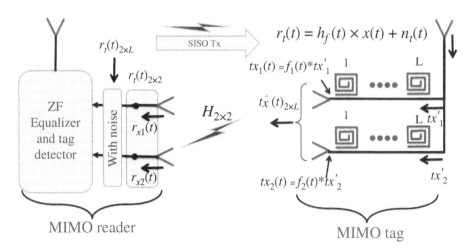

Figure 12.3   MIMO tag operation overview.

$$tx_1(t) = f_1(t) * tx_1'$$

$$= f_1(t) * \left(\frac{1}{\sqrt{2}} \left(h_f x(t) + n_t(t) \right) \right)$$

$$tx_2(t) = f_2(t) * tx_2'$$

$$= f_2(t) * \left(\frac{1}{\sqrt{2}} \left(h_f x(t) + n_t(t) \right) \right)$$

$$(12.2)$$

When the tags are located closer to the reader, the signal power is considerably higher than the noise added by the receiving antenna of the tag. Therefore, noise can be neglected and the above-mentioned expressions further simplify to (12.3). $S_m^{(1)}(t)$ and $S_m^{(2)}(t)$ are the resultant signals after tag modulation in branches 1 and 2, respectively, which are independent of the forward channel.

$$tx_1(t) \approx \frac{1}{\sqrt{2}} h_f \left(f_1(t) * x(t) \right)$$

$$\approx \frac{1}{\sqrt{2}} h_f(t) S_m^{(1)}(t)$$

$$tx_2(t) \approx \frac{1}{\sqrt{2}} h_f \left(f_2(t) * x(t) \right)$$

$$\approx \frac{1}{\sqrt{2}} h_f(t) S_m^{(2)}(t)$$

$$(12.3)$$

If the tag responses in (12.3) has L number of samples, then these two responses can be stacked to form a matrix $tx(t)_{2 \times L}$ as shown (12.4).

$$tx(t) = \begin{pmatrix} tx_1(t) \\ tx_2(t) \end{pmatrix}_{2 \times L} = \frac{1}{\sqrt{2}} h_f \begin{pmatrix} S_m^{(1)}(t) \\ S_m^{(2)}(t) \end{pmatrix}_{2 \times L} \qquad (12.4)$$

The received signals ($rx_1(t)$ and $rx_2(t)$) at the reader antenna array can be represented using a matrix as follows:

$$r_r(t) = \begin{pmatrix} rx_1(t) \\ rx_2(t) \end{pmatrix}_{2 \times L}$$

If the reverse channel is given by $hr_{2 \times 2}$, the received signal array at the RFID reader can be calculated using (12.5). $H_{2 \times 2}$ is the product of the forward and the reverse channels weighted by a factor of $\frac{1}{\sqrt{2}}$, and the noise matrix added by the both receiving antennas of the reader is given by $n_r(t)_{2 \times L}$.

$$r_r(t)_{2 \times L} = hr_{2 \times 2} \times tx(t)_{2 \times L} + n_r(t)_{2 \times L}$$

$$= \frac{1}{\sqrt{2}} h_f hr_{2 \times 2} \times \begin{pmatrix} S_m^{(1)}(t) \\ S_m^{(2)}(t) \end{pmatrix}_{2 \times L} + n_r(t)_{2 \times L}$$

$$= H_{2 \times 2} \times \begin{pmatrix} S_m^{(1)}(t) \\ S_m^{(2)}(t) \end{pmatrix}_{2 \times L} + n_r(t)_{2 \times L} \qquad (12.5)$$

Assuming the channel, $H_{2\times2}$ is known to the RFID reader, the estimated tag responses can be calculated using standard MIMO decomposing methods. In this section, two methods are presented, namely, zero forcing (ZF) equalizer and the minimum mean square error equalizer (MMSE) as presented in (12.6).

$$W_{ZF} = \left(H^{\#}H\right)^{-1}H^{\#}$$

$$W_{MMSE} = \left(H^{\#}H + N_0 I_{2\times2}\right)^{-1}H^{\#} \tag{12.6}$$

$H^{\#}$ is the Hermitian transpose of the channel matrix H and N_0 is the noise power available at the reader, which is calculated using SNR.

The ML-based tag detection technique derived in Section 12.4 assumes a Gaussian distribution for noise available at the reader. We have used MMSE equalizer as the decomposing technique in prior work [1]. The results are shown in Section 12.6.1 under method 1. Even though it leads to better signal to interference plus noise ratio (SINR) performance, in the process it makes noise distribution to be bimodal, hence it no longer can be treated as Gaussian. On the other hand, the noise produced after ZF method still follows a Gaussian distribution subjected to an amplified noise. Modeling the bimodal noise distribution to derive a likelihood-based detector makes the signal processing extremely complex. Therefore, only the ZF equalizer is used with the likelihood detector and the results show that even under this scenario the ML detection technique performs better than the performance reported in Ref. [1]. This method is named as method 2 and the results are presented in Section 12.6.2.

After applying the ZF equalizer, an estimated tag response for each branch can be obtained as shown in (12.7). $Y_{2\times L}(t)$ represents the estimated tag responses in each branch.

$$Y_{2\times L}(t) = W_{ZF} \times r_r(t)_{2\times L} \tag{12.7}$$

The estimated tag response in (12.7) is derived for time-domain-based signal samples. However, the above relationship will still be applicable if a unitary transformation such as Fourier transform is performed. As a result, the estimated tag responses in frequency domain can be calculated as follows:

$$Y_{2\times L}(f) = W_{ZF} \times R_r(f)_{2\times L} \tag{12.8}$$

An ML-based tag detection technique is derived, which is later applied on these estimated tag responses. Derivation of the tag detection technique is discussed in the following section.

12.3   TAG DETECTION IN MIMO

In this section, we derive an expression for the maximum likelihood (ML) function, to detect which resonator combination (notch filters) the signal has gone through. ZF equalizer derived in the previous section produces an estimated tag response for each branch in the MIMO tag. However, it could also amplify the noise during the decomposing. As a result, the output of the ZF equalizer can be expected to be noisy. The task is to find which resonator combination has the highest maximum likelihood, out of all the possibilities. For example, if each branch had N resonators, then there would be 2^N unique tag responses in each branch. Therefore, each estimated tag response should be compared with 2^N tag responses and the one with the highest likelihood is selected.

The signal model used is explained in what follows. If $\bar{S}_m$ is the mth tag response vector out of all the 2^N number of combinations and $\bar{\omega}$ is the noise vector available after ZF equalizer, then the estimated tag response vector, $\bar{y}$ is given by

$$\bar{y} = \bar{S}_m + \bar{\omega}$$

Due to I & Q demodulation, these signals are complex and they can be represented using real and imaginary components ($[.]_r$ and $[.]_i$) as follows:

$$\bar{y} = \bar{y}_r + j\bar{y}_i$$
$$\bar{S}_m = \bar{S}_{m,r} + j\bar{S}_{m,i}$$
$$\bar{\omega} = \bar{\omega}_r + j\bar{\omega}_i$$

Therefore, the received signal can be written as

$$\bar{y}_r = \bar{S}_{m,r} + \bar{\omega}_r$$
$$\bar{y}_i = \bar{S}_{m,i} + \bar{\omega}_i \tag{12.9}$$

Each noise sample in $\bar{\omega}$ is assumed to have an independent and identical distribution (i.i.d.) with zero mean and a variance of σ^2. Original noise added at the reader is assumed to be Gaussian due to the reader architecture. As pointed out in the previous section, ZF equalizer may only amplify the noise. As a result, the noise after the equalizer can still be treated as Gaussian. Therefore, it was assumed both the real and imaginary parts of each noise sample (ω_i) has a Gaussian distribution given by, $\omega_i \sim N(0, \sigma^2)$. Then for calculation purposes, the results can be vectorized as (12.10).

$$\bar{\omega}_r \sim N(\bar{0}, \sigma^2 I_N)$$
$$\bar{\omega}_i \sim N(\bar{0}, \sigma^2 I_N) \tag{12.10}$$

I_N is the identity matrix with dimensions of $N \times N$. A new real-valued vector, $\bar{y}_0$ is created using $\bar{y}_r$ and $\bar{y}_i$ as follows:

$$\bar{y}_0 = (\bar{y}_r, \bar{y}_i) \tag{12.11}$$

Mean and covariance of $\bar{y}_0$ can be calculated as follows:

$$E[\bar{y}_0] = \bar{\mu} = \begin{bmatrix} \bar{S}_{m,r}, \bar{S}_{m,i} \end{bmatrix}$$
$$\mathrm{Cov}[\bar{y}_0] = E\left[(\bar{y}_0 - \bar{\mu})^T (\bar{y}_0 - \bar{\mu}) \right]$$
$$= \sigma^2 I_{2N} \tag{12.12}$$

I_{2N} in (12.12) is the identity matrix with a dimension of $2N \times 2N$. Using the statistical properties calculated in (12.12), the distribution of the real vector $\bar{y}_0$ can be represented as follows:

$$\bar{y}_0 \sim N(\bar{\mu}, \sigma^2 I_{2N})$$

Using probability theory, the conditional probability of receiving $\bar{y}$ given that $\bar{S}_m$ has been transmitted can be derived as (12.13)

$$
\begin{aligned}
\Pr(\bar{y}_0 | \bar{S}_m) &= \frac{1}{2\pi \sqrt{|\mathrm{Cov}[\bar{y}_0]|}} \\
&\times \exp\left(-\frac{1}{2} \times (\bar{y}_0 - \bar{\mu})\mathrm{Cov}(\bar{y}_0)^{-1}(\bar{y}_0 - \bar{\mu})^T\right) \\
&= \frac{1}{2\pi\sigma} \exp\left(-\frac{1}{2\sigma^2}(\bar{y}_0 - \bar{\mu})(\bar{y}_0 - \bar{\mu})^T\right)
\end{aligned}
\tag{12.13}
$$

Equation (12.13) is evaluated for all the possible tag combinations, and the one with the highest probability is taken as the detector output. However, it can be seen that the detector can be further simplified to minimizing the $\exp(.)$ component. Therefore, the objective function of the detector can be represented as follows:

$$\max_{\bar{S}_m} \Pr(\bar{y}_0 | \bar{S}_m) = \min_{\bar{S}_m} \left((\bar{y}_0 - \bar{\mu})(\bar{y}_0 - \bar{\mu})^T\right) \tag{12.14}$$

Under the assumptions made for the proposed signal model, it can be proved that the optimum detector is the same as the minimum distance detector. However, $\bar{y}_0$ and $\bar{\mu}$ can be calculated using (12.11) and (12.12), respectively. In addition, this expression is valid for frequency-based samples as unitary transformations such as FFT do not change the statistical properties of the signals.

12.4 EXPERIMENTAL SETUP

An experiment was conducted to measure the MIMO tag response using an arbitrary waveform generator (AWG) and an oscilloscope with a high sampling rate as shown in Figure 12.4. However, antennas were replaced using cables as the sole purpose of this work is to verify the validity of the ML-based detection method.

Figure 12.5 shows the CST-generated tag response and the measured tag response for tag bits [1010]. It can be seen that they are closely matched.

An 8-bit MIMO tag was fabricated to have tag bits [1010] in the first branch and [0000] in the other. An AWG was used to generate the interrogating signal at 2.4 GHz and oscilloscope with 20 GSamples/s was used to capture the tag responses available at each branch. ML-based tag detection was performed only on the first branch for demonstrative purposes and the same technique can be applied to the second branch too. Table 12.1 shows the distance value obtained for each tag type after evaluating the expression in (12.14). It can be clearly seen that minimum distance occurs at the tag type [1010] as highlighted in Table 12.1. A comprehensive analysis was

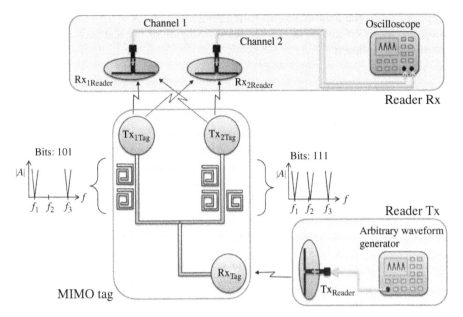

Figure 12.4   MIMO tag experiment.

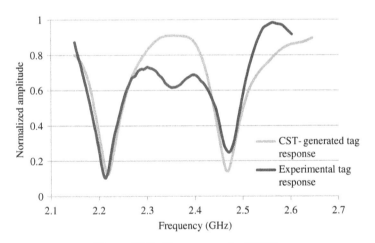

Figure 12.5   Tag response for [1010].

carried out using MATLAB simulations, as it is not feasible to take large number of experimental data.

12.5   SIMULATIONS

Simulations were carried out in two methods. In the first method, tag responses were generated in MATLAB using bandstop filters. The interrogating signal was generated

TABLE 12.1   An Example of a Table

Tag Type	Distance in (12.14)	Tag Type	Distance in (12.14)
[0000]	10.84	[1000]	4.73
[0001]	7.57	[1001]	4.12
[0010]	5.03	[1010]	0.56
[0011]	7.89	[1011]	2.80
[0100]	4.49	[1100]	1.62
[0101]	3.58	[1101]	2.75
[0110]	3.51	[1110]	2.61
[0111]	5.74	[1111]	5.29

using orthogonal frequency division multiplexing (OFDM) techniques. The MIMO decomposing technique explained earlier was implemented in MATLAB and the tag responses in each branch of the MIMO tag were estimated. In this method, tag detection was performed using a threshold-based valley detection technique applied on the power spectral density (PSD) of the estimated tag responses. This information was used to calculate the DER at different SNR levels.

In the second method, both the interrogating signals and the tag responses were generated using CST simulations. The MIMO decomposing technique and the tag detection technique described in the previous section were implemented in MATLAB. Finally, the DER at different SNR levels was calculated. First, the details about the first method are described next.

12.5.1   METHOD 1

The signal path from the RFID reader through the MIMO tag back to the reader was modeled in MATLAB using the baseband signal representation. As briefly explained earlier, in this method first the interrogating signal is generated using OFDM techniques. This was achieved using binary phase shift keying (BPSK)-modulated test bits. All 200 test bits were selected as "1." Other OFDM parameters used in MATLAB simulations are displayed in Table (12.2).

Then the resonators were emulated using bandstop filters in MATLAB and the filter attenuation was selected as 10 dB. Center frequencies of the bandstop filters were selected as shown in Table 12.2. Multiple resonators were emulated using cascaded bandstop filters. Then the tag responses were fed through a channel that is given by the product of the forward and reverse channels of the RFID system. In the MATLAB simulations, the channel is assumed to be known at the RFID reader. Finally, noise was added to the resultant signal according to the specified SNR.

Baseband signals received at the RFID reader was used for MIMO decomposing. Then ZF equalizer was used to decompose the received signal and obtain an estimate of the tag response in each branch. Then a threshold-based valley detector was used to identify the presence and absence of resonators, which will be used to read the encoded tag data bits. Finally, the DER was calculated for each tag detection technique at different SNR levels.

TABLE 12.2   **Simulation Parameters**

Parameter	Value
Number of BPSK modulated of test symbols	200
OFDM block size	25
Length of cyclic prefix	2
Number of FFT / IFFT points	25
Sampling frequency	600 MHz
OFDM signal bandwidth	300 MHz
Total bits encoded in the tag	6 bits
Bandstop filter attenuation	10 dB
Number of branches in the MIMO tag	2
Resonance frequency set (MSB to LSB)	[50, 150, 250, 350, 450, 550] MHz

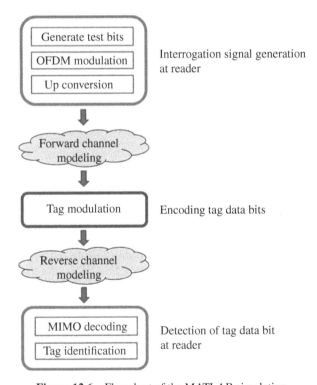

Figure 12.6   Flowchart of the MATLAB simulation.

Figure 12.6 illustrates the flowchart of the MATLAB simulation carried out. However, the calculation of tag responses using bandstop filters does not take into account any coupling between the two branches in the tag. Therefore, a more realistic simulation was performed using CST. In addition, valley detection technique used to identify the tag data bits is very primitive and can be further improved using advanced tag detection techniques. The second method explained in the following section rectifies these limitations.

12.5.2   Method 2

As explained in the previous section, method 2 uses CST to simulate tag responses, which takes into account any coupling between the two branches in the MIMO tag. The tag responses generated using CST simulations are more realistic than band-stop filter emulation performed in the previous method. Then the validity of the tag detection technique derived in *Section* 12.3 was verified using MATLAB.

The steps carried out in the simulation are given in Figure 12.7. First, an interrogating signal was generated to provide a flat frequency response in the 2.2–2.6 GHz frequency range. Four resonators were designed using CST with resonating frequencies as shown in Table 12.3. Then the combinations of resonators were placed beside a microstrip line to cover all possible tag IDs. One end of the microstrip line was fed with the interrogating signal and the tag responses were collected at the other end. These collected tag responses were saved in a lookup table for the algorithms to be used later.

Then the signal flow from the MIMO tag to the reader was modeled in MATLAB and the channel information was assumed to be available at the reader. ZF equalizer was used to decompose the tag responses in each branch of the MIMO tag. Then the tag detection technique derived in *Section* 12.3 was used to identify the encoded tag data bits. Finally, the detection error rate was calculated at different SNR levels. The results obtained in each simulation method are discussed next.

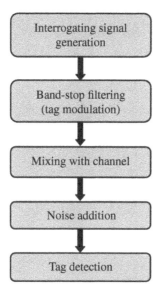

Figure 12.7   Flowchart of the MATLAB simulation.

TABLE 12.3   Simulation Parameters

Parameter	Value
Center Frequency	2.4 GHz
Total bits encoded in a tag	4 bits
Flat frequency response	400 MHz
Bandstop filter attenuation	10 dB
Guard band	50 MHz
Resonance frequency set (MSB to LSB)	[2.2, 2.3, 2.4, 2.5] GHz
Number of iterations	1,000,000

12.6   RESULTS

In this section, results obtained using the two simulation methods are presented. First, the results of the method using OFDM technique and bandstop filters are discussed.

12.6.1   Method 1

In the simulation, the test bits are selected as all ones and they are BPSK modulated. Then the BPSK-modulated signals are OFDM modulated to generate the interrogating signal in time domain as shown in Figure 12.8. When the test bits are taken as all ones, the resulting interrogating signal will be similar to having a train of impulses.

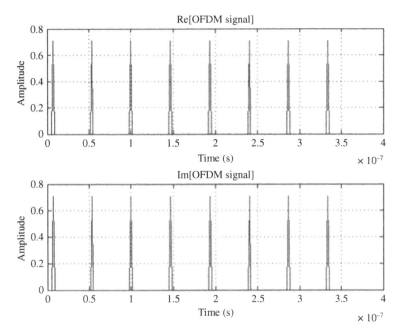

Figure 12.8   Interrogating signal in time domain.

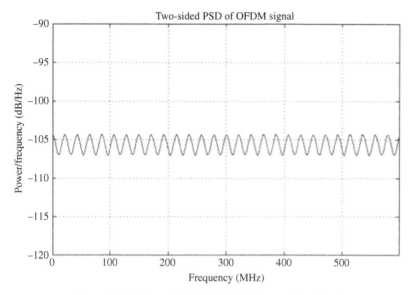

Figure 12.9 Two-sided PSD of the interrogating signal.

The two-sided PSD of the interrogating signal is shown in Figure 12.9. It can be considered that the signal has a flat frequency response throughout the signal bandwidth of 300 MHz.

This interrogating signal was then transmitted through a SISO channel to the tag and the time-domain representation of the received signal is shown in Figure 12.10.

Then the received signal at the tag is divided into two equal RF components and each component will travel surrounding spiral resonators. The spiral resonators were implemented as bandstop filters in MATLAB and Figure 12.11 shows the filter response of such a spiral resonator. Altogether three resonators were emulated at 47, 150, and 253 MHz. The presence of a resonator was represented as bit "1" while the absence as bit "0." As there are two components that are tag modulated independently, the considered prototype has a capacity of 6 bits.

Once each component reaches its transmitting antenna, it contains the frequency signature of all the spirals presented along the way. Figure 12.12 shows the two-sided PSDs of each signal (Tx1 and Tx2) after traveling via spiral resonators. Top graph in Figure 12.12 represents the case of having only two resonators at 47 and 253 MHz (corresponds to [101]) while the bottom represents having three resonators at 47, 150 and 253 MHz (corresponds to bits [111]). Presence and absence of the spiral resonators in each branch can clearly be observed.

The time-domain representation of the above two signals is shown in Figure 12.13. The noise introduced by the receiving tag antennas is visible at the two signals already.

Then the two transmitted signals were propagated via a 2×2 MIMO channel. Figure 12.14 shows the channel realizations for both forward and reverse channels.

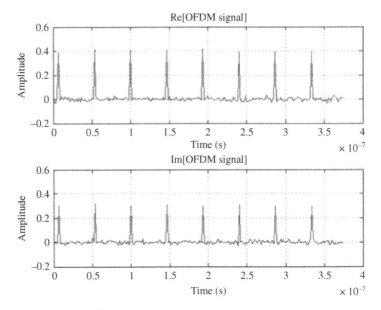

Figure 12.10 Received signal at the tag.

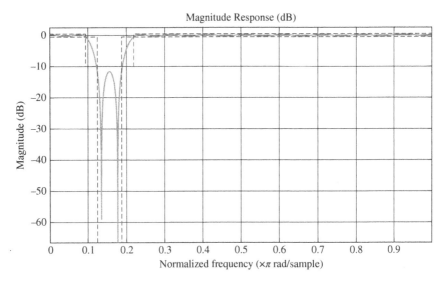

Figure 12.11 Filter response of a spiral resonator.

As a result of Rician distribution, the four MIMO channel realizations looked very similar. After being mixed with the channel, the received signals at the receiver antenna array are shown in Figure 12.15.

The received signal array were then decoded using MMSE equalizing method and the estimated transmitted signals (Tx1H and Tx2H) were obtained. Figures 12.16

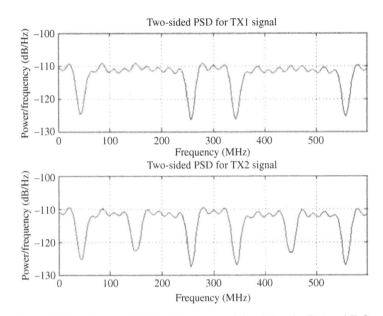

Figure 12.12 Two-sided PSD of the tag-modulated signals (Tx1 and Tx2).

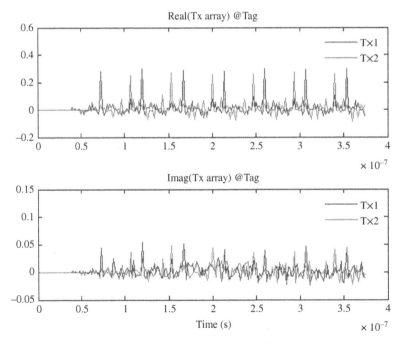

Figure 12.13 Tag-modulated signals (Tx1 and Tx2) in time domain.

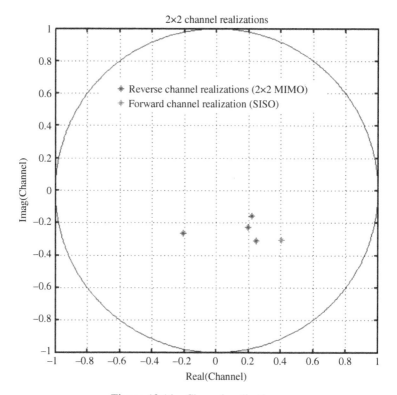

Figure 12.14   Channel realizations.

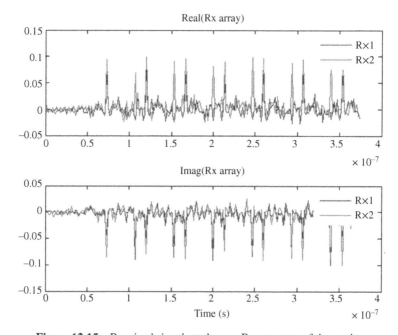

Figure 12.15   Received signals at the two Rx antennas of the reader.

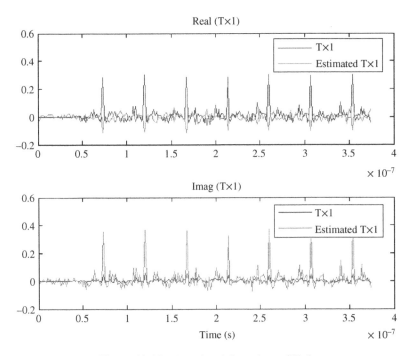

Figure 12.16   Actual and the estimated Tx1.

and 12.17 compare the actual and the estimated transmitted signals in time domain for Tx1 and Tx2, respectively. It can be observed that the estimates are very similar to the actual signals.

The two-sided PSDs of each of the estimated transmitted signals (Tx1 and Tx2) are shown in Figure 12.18. A separate algorithm was implemented to detect the presence and absence of the frequency dips at the resonating frequencies. Using the algorithm, estimated data bits were obtained and were compared with the data bits encoded in the tag:

The simulation was repeated for 100 times, and Figure 12.19 shows the combined tag response obtained for 100 iterations. It can be concluded that for different channel realizations, the performances are consistent.

Simulations are carried out to investigate the bit error rate (BER) performance under different signal-to-noise ratios (SNRs). SNR is defined as the signal-to-noise ratio at each of the receiving antenna array at the reader. Figure 12.20 shows the BER performance of the proposed system versus SNR and also a comparison to the theoretical BER of a traditional binary phase shift keying (BPSK) modulation scheme. Even though, the definition of the BERs in two schemes is different, it is interesting to learn that the simulated system performances closely follow the theoretical BER performance for a BPSK-modulated 2×2 MIMO system. In traditional BPSK modulation schemes, the bits are modulated into either raised cosine symbols or no signal at all based on the bit value. In the PSD, it is similar to be represented using the presence

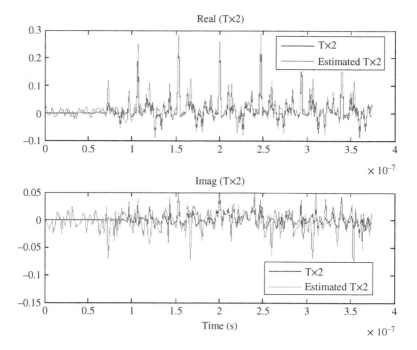

Figure 12.17   Actual and the estimated Tx2.

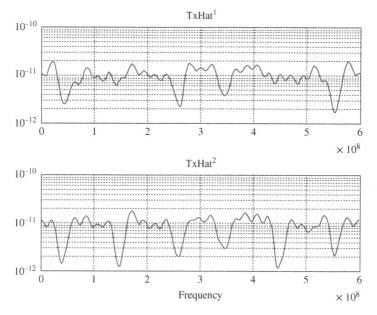

Figure 12.18   Combined tag response.

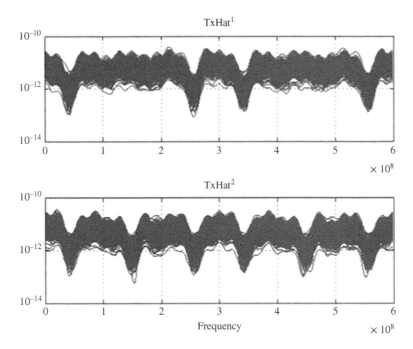

Figure 12.19 Combined tag response for 100 iterations.

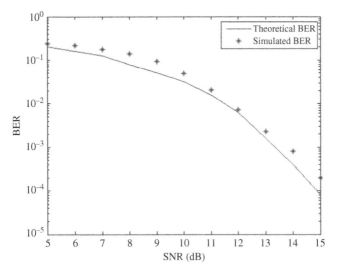

Figure 12.20 BER of the proposed system versus SNR.

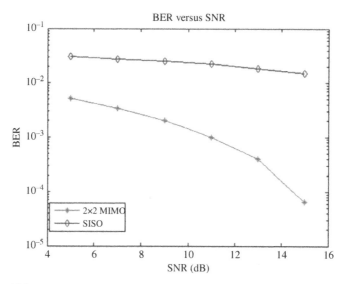

Figure 12.21   Noise performance of the proposed system versus SISO counterpart.

and absence of a valley. As a result, the two schemes can be compared and should display similar performance which they do.

Figure 12.21 shows a comparison between the proposed 2×2 MIMO system and the traditional SISO multiresonator-based chipless RFID system. In this simulation, the same data bits were encoded in both the branches introducing diversity. Compared with the traditional system, it is evident that the two branches in the MIMO multiresonator tag cause less errors due to the extra reliability in the proposed system. This extra reliability can also be seen differently as encoding more data bits in the tag with an acceptable reading accuracy. Apart from that, the diversity gain of the MIMO system is clearly visible compared to linear variation in the SISO system.

12.6.2   Method 2

The results obtained using both CST and MATLAB simulations are presented and discussed in this section. The resonator response simulation is similar to that of the SISO tag simulation. Figure 12.22 shows the resonator response of a branch when all four resonators are presented. It can be clearly seen that the four resonances at the designed frequencies 2.2, 2.3, 2.4, and 2.5 GHz.

The received signal at each receiving antenna of the reader is calculated using (12.9) under different noise power levels. Then the received signal array is decomposed using ZF decoder as shown in (12.7) to calculate the estimated tag responses in each branch. Finally, the likelihood-based detector derived in (12.14) is used to identify the tag data bits. Figure 12.23 compares the DER against SNR for the two methods.

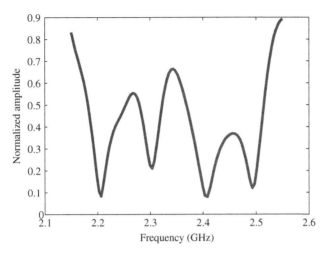

Figure 12.22 CST-generated tag response for a branch having [1111] tag bits.

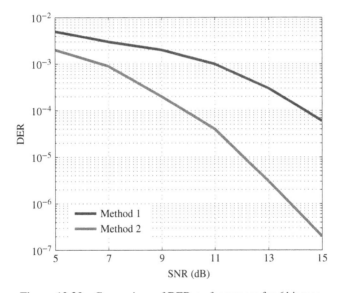

Figure 12.23 Comparison of DER performances for 6 bit tags.

It can be clearly seen that the likelihood-based detection technique (method 2) is performing better than the valley detection-based technique in method 1. For example, at SNR = 10dB, method 1 has a DER of 99.8%, while method 2 produces an accuracy of 99.99%. At higher SNR levels, both methods perform better for obvious reasons. It can be concluded that the likelihood-based detection method performs better than the valley detection method at all SNR levels. However, method 2 assumes perfect channel state information. In case if there are errors in channel estimation, it could magnify the noise amplitude with ZF decoder and performances could be degraded.

12.7 CONCLUSION

This chapter presented signal processing techniques for successfully detecting the tag bits of a MIMO-based chipless RFID system. First, ZF decomposing techniques were used on the received data array at the reader to estimate the tag response. Then two methods were used to detect the tag data bits encoded in the MIMO tag. The first method uses a threshold-based valley detection method. The other method uses the proposed ML detection technique to detect the tag bits in each branch of the MIMO chipless tag. An experiment was set up to test the ML detection technique and it was shown that the encoded tag data bits are identified successfully. In order to perform a comprehensive analysis, CST- and MATLAB-based simulations were performed. The results show that the proposed detection technique provides better detection error rate performance at different SNR values over traditional threshold-based detection. This benefit can be interpreted in two different metrics. First, it can be seen as an SNR gain over the existing threshold-based detection technique, which effectively increases the reading range. Second, the high accuracy in tag reading avoids multiple reading cycles, which yields an energy-efficient reading method.

Theoretically, higher number of branches can encode higher tag data bits within the same frequency band. However, the number of receiving antennas used in the reader should be always equal to or higher than the total branches in the MIMO tag to successfully decompose the signals. In addition, due to short distance the proposed setup operates under line-of-sight (LOS) MIMO. In LOS MIMO, the physical distances between antennas have to be maintained such that they will not form similar channel gains between transmitter and receiver antennas. As a result, the maximum number of branches allowed in the tag is limited. Main drawback of ML detection technique is computation complexity. It was demonstrated in *Chapter 11* that the proposed computationally feasible detection techniques reduce the complexity from exponential to linear order without compromising the tag reading accuracy.

REFERENCE

1. C. Divarathne and N. Karmakar, "MIMO based chipless RFID system," in *RFID-Technologies and Applications (RFID-TA), 2012 IEEE International Conference* on, Nov 2012, pp. 423–428.

13

CONCLUSION FOR PART II

Chapters 10 to *12* in Part II of the book has presented a number of chipless RFID tag detection techniques including two computationally feasible tag detection techniques. In addition, a multiple input multiple output (MIMO)-based chipless radiofrequency identification (RFID) system has been proposed, which is the first of its kind reported to the best of the knowledge of the author. This chapter provides a summary of the proposed techniques in Part II including the limitations of the proposed tag detection techniques and the potential applications. Finally, it presents a set of recommendations for chipless RFID tag detection.

13.1 SUMMARY OF THE PROPOSED TECHNIQUES IN PART II

Part II of this book presents advanced yet computationally feasible tag detection techniques for chipless RFID systems that are capable of improving the tag reading accuracy, reading range, and data bit capacity. Detection error rate of a number of likelihood-based detectors was presented and compared against the threshold-based detector used in existing chipless RFID systems. It is evident that all of the likelihood-based detectors perform better than the popular threshold-based detector. The performance improvement of the proposed tag detection techniques can be interpreted as an increased tag reading accuracy at a given signal-to-noise ratio (SNR) level. On the other hand, it can also be represented as an increment in the reading range while achieving a particular goal of reading accuracy. Therefore, the improved performance can be represented either as increased reading accuracy or reading range depending on the application requirement. In addition, two approaches

Advanced Chipless RFID: MIMO-Based Imaging at 60 GHz–ML Detection, First Edition.
Nemai Chandra Karmakar, Mohammad Zomorrodi, and Chamath Divarathne.
© 2016 John Wiley & Sons, Inc. Published 2016 by John Wiley & Sons, Inc.

have been taken to improve the tag data bits. First, the proposed tag detection techniques have allowed to remove the guard-band present in the frequency-domain tags. It has been shown that it allows to increase the data capacity by a factor up to 2. The second approach is to design a new MIMO chipless RFID tag and the relevant signal processing techniques. Theoretically, it can be proved that the tag data capacity can be improved by a factor of 2 or greater.

However, there is a common drawback of all the likelihood-based detection methods discussed so far. All these methods require higher computation complexity compared to the primitive detection techniques such as threshold-based detection. Computationally feasible two tag detection techniques have been proposed to reduce the computation complexity from exponential to linear in *Chapter 11*. It was found that the bit-by-bit detection method, which is a suboptimal detection method, performs successfully when a resonator guard-band is used in tag design. It was shown that the computation time has significantly dropped compared to exhaustive maximum likelihood detection methods without compromising the reading accuracy. In addition, a fully optimal trellis-tree-based Viterbi decoding technique has been introduced to reduce the computation complexity from exponential to linear order while achieving a similar reading accuracy to original likelihood detection techniques. *Chapter 12* presents signal processing techniques for successfully detecting the tag bits of a MIMO-based chipless RFID system. First, zero forcing decomposing techniques were used on the received data array at the reader to estimate the tag response. Then two methods were used to detect the tag data bits encoded in the MIMO tag. In the first method [1], tag responses were generated in MATLAB using bandstop filters. The first method uses a threshold-based valley detection method. The other method uses the proposed ML detection technique to detect the tag bits in each branch of the MIMO chipless tag. An experiment was setup to test the ML detection technique and it was shown that the encoded tag data bits were identified successfully. A comprehensive analysis was performed using a CST- and MATLAB-based simulation. The results show that the proposed detection technique provides better detection error rate performance at different SNR values over traditional threshold-based detection. This benefit can be interpreted in two different metrics. First, it can be seen as an SNR gain over the existing threshold-based detection technique, which effectively increases the reading range. Second, the high accuracy in tag reading avoids multiple reading cycles, which yields an energy-efficient reading method.

Theoretically, higher number of branches can encode higher tag data bits within the same frequency band. However, the number of receiving antennas used in the reader should be always equal to or higher than the total branches in the MIMO tag to successfully decompose the signals. In addition, due to short distance, the proposed setup operates under line-of-sight (LOS) MIMO. In LOS MIMO, the physical distances between antennas have to be maintained such that they will not form similar channel gains between transmitter and receiver antennas. As a result, the maximum number of branches allowed in the tag is limited. Main drawback of ML detection technique is the computation complexity. It was demonstrated in *Chapter 11* that the proposed computationally feasible detection techniques reduce the complexity from exponential to linear order without compromising the tag reading accuracy.

After analyzing the simulations, it is noteworthy to pinpoint that, even though there are only two transmitting branches present in the RFID tag considered, it is theoretically possible to add more branches and still recover the transmitted signals given that the number of receiving antennas in the reader is larger than or equal to the number of transmitting branches in the tag. Hence, without increasing the bandwidth, the bit capacity can be further increased using the same frequency resonators compared with having only one branch at the tag. However, it is required to evaluate the effect of mutual coupling between antennas with higher number of transmitting branches in the tag.

In the proposed RFID tag, there is only one receiving antenna through which the received signal will be divided into two equal components. The proposed concept can be extended to having a dedicated receiving antenna for each component, hence increasing the effective SNR at each branch. Therefore, the performances can be further improved with the usage of multiple dedicated transmitting and receiving antennas on the tag. In addition, the concept can be further extended to multiple tag detection if each branch is considered as a separate tag. Furthermore, the use of I/Q modulation/demodulation allows an extra degree of freedom to increase the bit capacity. Since the baseband signal considered is complex, it is possible to have asymmetric frequency response in positive and negative frequencies. Therefore, the eligible frequency band in the passband centered around the RF carrier doubles, allowing more resonators to be placed in the tag, without increasing the sampling rate of the ADC at the receiving end of the reader. After analyzing the above results, it can be concluded that MIMO is a competitive candidate for improving reliability or the bit capacity of a resonator-based chipless RFID system.

It was found that the proposed tag detection techniques for single input single output (SISO) systems provide significantly higher tag reading accuracy over the existing threshold-based detector. In addition, they are capable of operating without a guard-band, which makes the tag data bit capacity to be doubled without compromising the reading accuracy. Moreover, the effective SNR gain provided by the proposed techniques can be represented as increasing tag reading range. All these benefits are achieved without compromising the low computation complexity. The MIMO tag with two branches is capable of encoding up to four times the total bits stored in existing SISO tags. Due to highly reliable tag detection techniques, chipless RFID tag readers do not need to read the same tag multiple times unlike the existing readers. This introduces the new ONE TIME tag reading philosophy.

13.2 LIMITATIONS OF THE PROPOSED SYSTEM

However, the performance of the proposed tag detection method could be limited by few factors. One of the main factors is the fabrication defects such as the dielectric constant of the substrate and the precision of the line widths. Due to these inaccuracies in tag design, two tags with the same tag data bits could have slightly different tag responses. These imperfections could affect the successful tag detection rate. In addition, when the tags are fabricated on paper, the resonance level is less compared to that on substrates. As a result, it could be more susceptible to noise conditions, which can cause the detection error rate to be increased.

The tag detection technique presented in System Model IV involves estimating the channel and then using the estimated channel for tag detection. Due to various conditions such as interference and object movement, the channel may change suddenly. The proposed tag detection technique assumes a slowly varying channel for the interrogation period, which is in the order of few hundreds of milliseconds. The sudden changes of the channel conditions introduce an error in channel estimation. This error can cause the detection error rate to be increased.

The proposed tag detection techniques perform well when there is only one tag in the vicinity of the reader interrogation zone. If there are multiple tags inside the interrogation zone, the responses from other tags interfere with the current tag of interest. In order to eliminate this interference, the channel realizations from all the tags to the reader should be known. However, obtaining these channel state information is very difficult as the positions of the channel are unknown and there is no feasible way to estimate the channel from each of those tags to the reader. If the tag positions are known, the reader can interrogate by beamforming only one tag at a time and record the tag response and estimate the channel as in one tag situation.

The tag detection techniques proposed in Part II require extra computational power compared to low-spec microcontrollers used in some of the chipless RFID readers. As a result, unless the existing hardware performance is already enough, there is a hardware upgrade for the new detection techniques to be worked. However, it can be seen that this hardware upgrade is feasible with single board computers as discussed in Section 13.4.

Even under the limitations presented earlier, the proposed smart tag detection techniques are expected to increase the data bit capacity in chipless RFID tags that can be detected at a higher success rate and be detected farther away from the reader. These advances in knowledge are expected to produce commercialized chipless RFID systems in future.

13.3 POTENTIAL APPLICATIONS

The proposed tag detection techniques can be used in a number of potential applications. The most favorable would be conveyer belt applications when a tag is either printed directly on the product or the already printed tag is stuck on the product. On a conveyer belt, the items to be identified can be controlled to appear one after the other. This avoids multiple tags being illuminated by the reader at the same time, hence interference-limited tag reading can be performed. In addition, this controlled item movement is important to avoid disorientation of items as tag reading is orientation sensitive. Some of these applications can be found in production lines in manufacturing industries, packaging, pharmaceuticals, and airport luggage tracking and handling.

Another potential application is to identify counterfeit bank notes. The tag will be printed on the polymer note with an invisible conductive ink using an inkjet or laser printer. A chipless RFID reader with the proposed tag detection techniques can be used to interrogate the banknotes and based on the detected data bits, counterfeit notes can be identified. Reserve bank of Australia is the world leader in printing polymer-based bank notes. They are currently working with the main author's research group to investigate the feasibility of implementing this technology to the bank notes.

Smart library is a concept that has been proposed for some time now. A smart library automates several day-to-day tasks with the use of RFID systems. The most popular task is the lending, where user picks a book and can check out using the RFID readers available at a self-checkout desk. This has already been realized on several occasions and the proposed detection techniques can help reliably perform several other tasks such as receiving new stocks, carrying out inventory checks, checking for misfiled items. For example, checking for misfiled items can be performed by scanning the book with a tag printed on it using a handheld chipless RFID reader.

It has been discussed about applying chipless RFID in retail market for a decade or so. The cost of fabricating a chipless RFID tag is less than a fraction of a cent, which makes it an ideal technology for tagging low-cost items (US $1 bread) in the retail market such as supermarket. Current limitations for the deployment are the low number of tag data bit and being unable to read multiple tags simultaneously. The proposed techniques help to double the data capacity by removing the guard-band and with improved tag detection techniques. So these techniques help to move one step closer to actual deployment.

There are few other areas where the proposed chipless RFID systems can be deployed. One of them is vehicle tracking where a tag is placed on the windscreen of the vehicle and readers are mounted at the entrance to the carpark. The authors collaborated with a ski center already performed a trial to check the feasibility of tracking the incoming and outgoing cars to the car park of the ski center. Tagging and tracking of individual components used in safety critical applications is another arena where the proposed chipless RFID system can be utilized. Some recommendations and open issues are presented next.

13.4   FUTURE WORK AND OPEN ISSUES

The verification of the proposed tag detection techniques was performed as a postprocessing exercise using MATLAB. Implementation of the detection techniques as a firmware is a significant step in producing commercialized chipless RFID readers having extra benefits summarized in the previous section. Table 13.1 outlines the specification of the latest Raspberry Pi 2 Model B that costs less than US $45 off the shelf.

Single board computers are becoming powerful than ever and are capable of loading advanced operating systems such as Windows or Linux. The quad-core processor and 1 GB of RAM make it possible to run powerful signal processing applications

TABLE 13.1 Technical Specifications of Raspberry Pi 2 Model B

Broadcom BCM2836 Arm7 quad-core processor running at 900 MHz
1 GB RAM
40-Pin extended GPIO
Micro-SD port for loading your operating system and storing data
Micro-USB power source
4 ×USB two ports
Four-pole stereo output and composite video port
Full-size HDMI
DSI display port for connecting the Raspberry Pi touch screen display

such as MATLAB. However, there are open-source software tools such as Octave, which is highly compatible with running MATLAB codes. In addition, extended 40-pin general-purpose input/output (GPIO) allows to capture signals for real-time data processing. These advances in technology and the cheap price have enabled single board computers to be a potential candidate for preparing portable chipless RFID readers with advanced tag detection techniques.

The second part of the work presented in Part II focuses on tag detection techniques for MIMO-based chipless RFID systems. The proposed detection techniques require accurate channel state information. More advanced signal processing techniques are required to be developed that perform when perfect channel state information is not available. In addition, the detection techniques were derived based on the assumption that the noise has identically independent Gaussian distributed samples. Even though the results produced are improved, it might be worth investigating a novel model to represent noise encountered in the proposed system. Moreover, the interrogating signal used is constructed based on having equal power across the frequency band of interest. However, once channel state information is available, the interrogating signal shape can be optimized to improve the tag reading accuracy.

With the proposed tag detection techniques, it is believed that the challenges for commercializing chipless RFID systems will be successfully overcome. As a result, chipless RFID can be made to be the future of barcode as the researchers predicted about a decade ago.

REFERENCE

1. C. Divarathne and N. Karmakar, "MIMO based chipless RFID system," in *RFID-Technologies and Applications (RFID-TA), 2012 IEEE International Conference on*, Nov 2012, pp. 423–428.

INDEX

Advanced Chipless RFID: MIMO-Based Imaging at 60 GHz–ML Detection, First Edition.
Nemai Chandra Karmakar, Mohammad Zomorrodi, and Chamath Divarathne.
© 2016 John Wiley & Sons, Inc. Published 2016 by John Wiley & Sons, Inc.

WILEY SERIES IN MICROWAVE AND OPTICAL ENGINEERING

Kai Chang, Series Editor

DIODE LASERS AND PHOTONIC INTEGRATED CIRCUITS,
Second Edition ● *Larry Coldren, Scott Corzine, and Milan Masanovic*

EM DETECTION OF CONCEALED TARGETS ● *David J. Daniels*

RADIO FREQUENCY CIRCUIT DESIGN ● *W. Alan Davis and Krishna Agarwal*

RADIO FREQUENCY CIRCUIT DESIGN, Second Edition ● *W. Alan Davis*

FUNDAMENTALS OF OPTICAL FIBER SENSORS ● *Zujie Fang, Ken K. Chin, Ronghui Qu, and Haiwen Cai*

MULTICONDUCTOR TRANSMISSION-LINE STRUCTURES: MODAL ANALYSIS TECHNIQUES ● *J. A. Brandão Faria*

PHASED ARRAY-BASED SYSTEMS AND APPLICATIONS ● *Nick Fourikis*

SOLAR CELLS AND THEIR APPLICATIONS, Second Edition ● *Lewis M. Fraas and Larry D. Partain*

FUNDAMENTALS OF MICROWAVE TRANSMISSION LINES ● *Jon C. Freeman*

OPTICAL SEMICONDUCTOR DEVICES ● *Mitsuo Fukuda*

MICROSTRIP CIRCUITS ● *Fred Gardiol*

HIGH-SPEED VLSI INTERCONNECTIONS, Second Edition ● *Ashok K. Goel*

FUNDAMENTALS OF WAVELETS: THEORY, ALGORITHMS, AND APPLICATIONS, Second Edition ● *Jaideva C. Goswami and Andrew K. Chan*

HIGH-FREQUENCY ANALOG INTEGRATED CIRCUIT DESIGN
● *Ravender Goyal (ed.)*

RF AND MICROWAVE TRANSMITTER DESIGN ● *Andrei Grebennikov*

ANALYSIS AND DESIGN OF INTEGRATED CIRCUIT ANTENNA MODULES ● *K. C. Gupta and Peter S. Hall*

PHASED ARRAY ANTENNAS, Second Edition ● *R. C. Hansen*

STRIPLINE CIRCULATORS ● *Joseph Helszajn*

THE STRIPLINE CIRCULATOR: THEORY AND PRACTICE ● *Joseph Helszajn*

LOCALIZED WAVES ● *Hugo E. Hernández-Figueroa, Michel Zamboni-Rached, and Erasmo Recami (eds.)*

MICROSTRIP FILTERS FOR RF/MICROWAVE APPLICATIONS,
Second Edition ● *Jia-Sheng Hong*

MICROWAVE APPROACH TO HIGHLY IRREGULAR FIBER OPTICS
● *Huang Hung-Chia*

METAMATERIALS WITH NEGATIVE PARAMETERS: THEORY, DESIGN, AND MICROWAVE APPLICATIONS ● *Ricardo Marqués, Ferran Martín, and Mario Sorolla*

OPTOELECTRONIC PACKAGING ● *A. R. Mickelson, N. R. Basavanhally, and Y. C. Lee (eds.)*

OPTICAL CHARACTER RECOGNITION ● *Shunji Mori, Hirobumi Nishida, and Hiromitsu Yamada*

ANTENNAS FOR RADAR AND COMMUNICATIONS: A POLARIMETRIC APPROACH ● *Harold Mott*

INTEGRATED ACTIVE ANTENNAS AND SPATIAL POWER COMBINING ● *Julio A. Navarro and Kai Chang*

ANALYSIS METHODS FOR RF, MICROWAVE, AND MILLIMETER WAVE PLANAR TRANSMISSION LINE STRUCTURES ● *Cam Nguyen*

LASER DIODES AND THEIR APPLICATIONS TO COMMUNICATIONS AND INFORMATION PROCESSING ● *Takahiro Numai*

FREQUENCY CONTROL OF SEMICONDUCTOR LASERS ● *Motoichi Ohtsu (ed.)*

INVERSE SYNTHETIC APERTURE RADAR IMAGING WITH MATLAB ALGORITHMS ● *Caner Özdemir*

SILICA OPTICAL FIBER TECHNOLOGY FOR DEVICE AND COMPONENTS: DESIGN, FABRICATION, AND INTERNATIONAL STANDARDS ● *Un-Chul Paek and Kyunghwan Oh*

WAVELETS IN ELECTROMAGNETICS AND DEVICE MODELING ● *George W. Pan*

OPTICAL SWITCHING ● *Georgios Papadimitriou, Chrisoula Papazoglou, and Andreas S. Pomportsis*

MICROWAVE IMAGING ● *Matteo Pastorino*

ANALYSIS OF MULTICONDUCTOR TRANSMISSION LINES ● *Clayton R. Paul*

INTRODUCTION TO ELECTROMAGNETIC COMPATIBILITY, Second Edition ● *Clayton R. Paul*

ADAPTIVE OPTICS FOR VISION SCIENCE: PRINCIPLES, PRACTICES, DESIGN AND APPLICATIONS ● *Jason Porter, Hope Queener, Julianna Lin, Karen Thorn, and Abdul Awwal (eds.)*

ELECTROMAGNETIC OPTIMIZATION BY GENETIC ALGORITHMS ● *Yahya Rahmat-Samii and Eric Michielssen (eds.)*

PLANAR ANTENNAS FOR WIRELESS COMMUNICATIONS • *Kin-Lu Wong*

FREQUENCY SELECTIVE SURFACE AND GRID ARRAY • *T. K. Wu (ed.)*

PHOTONIC SENSING: PRINCIPLES AND APPLICATIONS FOR SAFETY AND SECURITY MONITORING • *Gaozhi Xiao and Wojtek J. Bock*

ACTIVE AND QUASI-OPTICAL ARRAYS FOR SOLID-STATE POWER COMBINING • *Robert A. York and Zoya B. Popoviαć (eds.)*

OPTICAL SIGNAL PROCESSING, COMPUTING AND NEURAL NETWORKS • *Francis T. S. Yu and Suganda Jutamulia*

ELECTROMAGNETIC SIMULATION TECHNIQUES BASED ON THE FDTD METHOD • *Wenhua Yu, Xiaoling Yang, Yongjun Liu, and Raj Mittra*

SiGe, GaAs, AND InP HETEROJUNCTION BIPOLAR TRANSISTORS • *Jiann Yuan*

PARALLEL SOLUTION OF INTEGRAL EQUATION-BASED EM PROBLEMS • *Yu Zhang and Tapan K. Sarkar*

ELECTRODYNAMICS OF SOLIDS AND MICROWAVE SUPERCONDUCTIVITY • *Shu-Ang Zhou*

MICROWAVE BANDPASS FILTERS FOR WIDEBAND COMMUNICATIONS • *Lei Zhu, Sheng Sun, and Rui Li*

FUNDAMENTALS OF MICROWAVE PHOTONICS • *Vincent Jude Urick Jr., Jason Dwight McKinney, and Keith Jake Williams*

RADIO-FREQUENCY INTEGRATED-CIRCUIT ENGINEERING • *Cam Nguyen*

ARTIFICIAL TRANSMISSION LINES FOR RF AND MICROWAVE APPLICATIONS • *Ferran Martín*

PASSIVE MACROMODELING • *Stefano Grivet-Talocia and Bjørn Gustavsen*